Precision Strike Warfare and International Intervention

This book explores whether the new capabilities made possible by precision strike technologies are reshaping approaches to international intervention.

Since the end of the Cold War, US technological superiority has led to a more proactive and, some would argue, high-risk approach to international military intervention. New technologies, including the capacity to mount precision military strikes from high-level bombing campaigns and, more recently, the selective targeting of individuals from unmanned aerial vehicles (UAVs) have facilitated air campaigns, supported by Special Forces, without the commitment of large numbers of troops on the ground.

Such campaigns include, for example, NATO's high-level aerial bombardment of Milošević's forces in Kosovo in 1999 and of Gaddafi's in Libya in 2011, and the US operation involving Special Forces against Osama Bin Laden. The development of UAVs and electronic data intercept technologies has further expanded the potential scope of interventions, for example against Islamic militants in the tribal areas of Pakistan.

This volume examines three key and interrelated dimensions of these new precision strike capabilities: (1) the strategic and foreign policy drivers and consequences; (2) the legal and moral implications of the new capabilities; and (3) the implications for decision making at the strategic, operational, and tactical levels.

This book will be of much interest to students of war and technology, air power, international intervention, security studies, and IR.

Mike Aaronson is the Executive Director of cii – the Centre for International Intervention – at the University of Surrey. He is a former chief executive of Save the Children UK and was previously a British diplomat.

Wali Aslam is Lecturer in International Relations at the University of Bath. He is the author of *The United States and Great Power Responsibility in International Society: Drones, rendition and invasion* (Routledge, 2013).

Tom Dyson is Senior Lecturer in International Relations at Royal Holloway College, University of London. He is the author of *European Defence Cooperation in EU Law and IR Theory* (2013, with Theodore Konstadinides).

Regina E. Rauxloh is Senior Lecturer at the Southampton School of Law, Southampton University. She is the author of *Plea Bargaining in National and International Law* (Routledge, 2012).

Routledge Global Security Studies

Series Editors: Aaron Karp and Regina Karp

Nuclear Proliferation and International Security
Edited by Morten Bremer Maerli and Sverre Lodgaard

Global Insurgency and the Future of Armed Conflict
Debating fourth-generation warfare
Terry Terriff, Aaron Karp and Regina Karp

Terrorism and Weapons of Mass Destruction
Responding to the challenge
Edited by Ian Bellany

Globalization and WMD Proliferation
Terrorism, transnational networks, and international security
Edited by James A. Russell and Jim J. Wirtz

Power Shifts, Strategy, and War
Declining states and international conflict
Dong Sun Lee

Energy Security and Global Politics
The militarization of resource management
Edited by Daniel Moran and James A. Russell

US Nuclear Weapons Policy after the Cold War
Russians, 'rogues' and domestic division
Nick Ritchie

Security and Post-Conflict Reconstruction
Dealing with fighters in the aftermath of war
Edited Robert Muggah

Network Centric Warfare and Coalition Operations
The new military operating system
Paul T. Mitchell

American Foreign Policy and the Politics of Fear
Threat inflation since 9/11
Edited by A. Trevor Thrall and Jane K. Cramer

Risk, Global Governance and Security
The other war on terror
Yee-Kuang Heng and Kenneth McDonagh

Nuclear Weapons and Cooperative Security in the 21st Century
The new disorder
Stephen J. Cimbala

Political Economy and Grand Strategy
A neoclassical realist view
Mark R. Brawley

Iran and Nuclear Weapons
Protracted conflict and proliferation
Saira Khan

US Strategy in Africa
AFRICOM, terrorism and security challenges
Edited by David J. Francis

Great Powers and Strategic Stability in the 21st Century
Competing visions of world order
Edited by Graeme P. Herd

The Globalisation of NATO
Intervention, security and identity
Veronica M. Kitchen

International Conflict in the Asia-Pacific
Patterns, consequences and management
Jacob Bercovitch and Mikio Oishi

Nuclear Proliferation and International Order
Challenges to the Non-Proliferation Treaty
Edited by Olav Njølstad

Nuclear Disarmament and Non-Proliferation
Towards a nuclear-weapon-free world?
Sverre Lodgaard

Nuclear Energy and Global Governance
Ensuring safety, security and non-proliferation
Trevor Findlay

Unipolarity and World Politics
A theory and its implications
Birthe Hansen

Disarmament Diplomacy and Human Security
Regimes, norms and moral progress in international relations
Denise Garcia

Causes and Consequences of Nuclear Proliferation
Edited by Robert Rauchhaus, Matthew Kroenig and Erik Gartzke

Why Did the United States Invade Iraq?
Edited by Jane K. Cramer and A. Trevor Thrall

Regional Powers and Security Orders
A theoretical framework
Edited by Robert Stewart-Ingersoll and Derrick Frazier

A Perpetual Menace
Nuclear weapons and international order
William Walker

Iran's Nuclear Programme
Strategic implications
Joachim Krause

Arms Control and Missile Proliferation in the Middle East
Edited by Bernd Kubbig

The National Politics of Nuclear Power
Economics, security and governance
Benjamin Sovacool and Scott Valentine

Arms Controls in the 21st Century
Between coercion and cooperation
Edited by Oliver Meier and Christopher Daase

Reconceptualising Deterrence
Nudging toward rationality in Middle Eastern rivalries
Elli Lieberman

Psychology, Strategy and Conflict
Perceptions of insecurity in International Relations
Edited by James W. Davis

Nuclear Terrorism and Global Security
The challenge of phasing out highly enriched uranium
Edited by Alan J. Kuperman

Ballistic Missile Defence and US National Security Policy
Normalisation and acceptance after the Cold War
Andrew Futter

Economic Statecraft and Foreign Policy
Sanctions, incentives and target state calculations
Jean-Marc F. Blanchard and Norrin M. Ripsman

Technology Transfers and Non-Proliferation
Between control and cooperation
Edited by Oliver Meier

Northern Security and Global Politics
Nordic-Baltic strategic influence in a post-unipolar world
Edited by Ann-Sofie Dahl and Pauli Järvenpää

Geopolitics and Security in the Arctic
Regional developments in a global world
Edited by Rolf Tamnes and Kristine Offerdal

Precision Strike Warfare and International Intervention
Strategic, ethico-legal, and decisional implications
Edited by Mike Aaronson, Wali Aslam, Tom Dyson, and Regina Rauxloh

Precision Strike Warfare and International Intervention

Strategic, ethico-legal, and decisional implications

Edited by Mike Aaronson, Wali Aslam, Tom Dyson, and Regina Rauxloh

Routledge
Taylor & Francis Group
LONDON AND NEW YORK

First published 2015
by Routledge

2 Park Square, Milton Park, Abingdon, Oxon OX14 4RN
711 Third Avenue, New York, NY 10017, USA

Routledge is an imprint of the Taylor & Francis Group, an informa business

First issued in paperback 2016

British Library Cataloguing-in-Publication Data
A catalogue record for this book is available from the British Library

Library of Congress Cataloging-in-Publication Data
Precision strike technology and international intervention: strategic, legal, and moral implications/edited by Mike Aaronson, Wali Aslam, Tom Dyson, and Regina Rauxloh.
pages cm – (Routledge global security studies)
Includes bibliographical references and index.
1. Air warfare–Case studies. 2. Air warfare–Moral and ethical aspects. 3. Air weapons–History–21st century. 4. Drone aircraft–History–21st century. 5. Drone aircraft–Moral and ethical aspects. 6. Intervention (International law). 7. Security, International. I. Aaronson, Mike, editor of compilation. II. Dyson, Tom, editor of compilation. III. Aslam, Wali, editor of compilation. IV. Rauxloh, Regina, editor of compilation.
UG630.P67 2013
358.4'24–dc23 2014007712

ISBN: 978-0-415-73020-4 (hbk)
ISBN: 978-1-138-20068-5 (pbk)

Typeset in Baskerville
by Wearset Ltd, Boldon, Tyne and Wear

This book is dedicated to Lynne Millward, whose inspiration helped shape this project, but who sadly did not live to see it realised.

Contents

Notes on contributors

Mike Aaronson, from 1995 to 2005 was Director General (chief executive) of Save the Children UK, and from 1988 to 1995 was the charity's Overseas Director. Mike is a founder member, and from 2001 to 2008 was Chair of the Board, of the Centre for Humanitarian Dialogue. He is an Honorary Fellow of Nuffield College, Oxford. In September 2008 he was appointed an Honorary Visiting Professor and in May 2011 a Professorial Research Fellow in the Department of Politics at the University of Surrey, where he is also Executive Director of cii – the Centre for International Intervention. He has been a Senior Adviser to NATO on the political/military aspects of NATO transformation, and has lectured at the UK Defence Academy on civil/military collaboration in peace support operations.

Laurence Alison is Director of the Centre for Critical and Major Incident Research, at the Institute of Psychology, Health and Society at the University of Liverpool. Professor Alison is a member of the University of Liverpool's Institute for Risk and Uncertainty and Co-Director of the Hydra Foundation. His research concentrates on two domains: advanced suspect and detainee interviewing and strategic high-risk decision making. He advises the courts, emergency services, and enforcement agencies, the security and defence services.

Wali Aslam is a Lecturer in International Relations at Bath University. Having attained a BA and an MA degree in Pakistan, he attended the University of Leeds for another MA and a PhD – both of which were in International Studies. Wali has previously held teaching positions at Brunel University, Forman Christian University (Lahore, Pakistan), and the University of Glasgow. Wali is the author of *The United States and Great Power Responsibility in International Society: Drones, Rendition and Invasion*, published by Routledge in 2013.

Adrian Banks is a Lecturer in the School of Psychology, University of Surrey. He is a cognitive scientist and chartered occupational psychologist with expertise in decision making, reasoning, and human factors.

The aim of his research is to understand and improve reasoning and decision making. His studies look at the interaction of intuitive and analytic thought and the nature of expertise. This work involves behavioural experiments, electrophysiology, and computer simulation. He has recently worked on the Independent Advisory Panel of the Defence Maritime Logistics School at HMS *Raleigh*.

André Barrinha is a Lecturer in Politics and International Relations at Canterbury Christ Church University and a Researcher at the Centre for Social Studies. He holds a PhD in International Relations from the University of Kent, Canterbury and an undergraduate degree in the same subject from the University of Coimbra. From 2004 to 2006 he worked at the Institute of Strategic and International Studies, Lisbon. His mains research interests are within the fields of Critical Security Studies, European security, Turkish Foreign Policy and International Relations Theory.

Luis da Vinha is an Assistant Professor at Valley City State University (USA). He holds a PhD in International Relations from the University of Coimbra (Portugal). Luis is also a guest researcher at the Research Centre of Geography and Planning of the University of Minho (Portugal). He works on foreign policy decision-making, US foreign policy, and geographic mental maps. Previously, Luis was the Head of the Geographic Information Bureau of the City of Coimbra and the Local Dissemination Manager for the European project CIVITAS Modern.

Tjerk de Greef is a Human Factors Engineer whose research interests include symbiotic man-machine interaction, requirement analysis, and rapid prototyping. In his research he follows the joint cognitive systems paradigm shift from automation extending human capabilities to automation partnering with the human leading to successful prototypes for the Royal Netherlands Navy and the Netherlands Urban Search and Rescue team. He is a part-time research fellow at the Department of Interactive Intelligence at Delft University of Technology and a research associate at the Oxford Institute for Ethics, Law and Armed Conflict. In addition, Tjerk also works as a Program Manager and Scientific Secretary for the ICT Research Program COMMIT.

Mandeep K. Dhami is Professor of Decision Psychology at Middlesex University. She has held academic posts at the University of Surrey, University of Cambridge, University of Victoria (Canada), University of Maryland (USA), and the Max Planck Institute for Human Development (Germany). Mandeep has also been a Principal Social Psychologist at DSTL (MoD). Her research focuses on human judgment, decision making and choice; understanding and communicating uncertainty; and risk perception and risk taking. She has examined these issues primarily in the criminal justice domain. Mandeep has provided advice and consultancy to government organisations in the UK and abroad (including the USA, Australia, Canada, and South Korea). She is Fellow of the Society for the Psychological Study of Social Issues (Division 9 of the American Psychological Association). Mandeep has published over 80 papers and is lead editor of *Judgment and Decision Making as a Skill*, published by Cambridge University Press in 2011.

Tom Dyson is Senior Lecturer in International Relations at the Department of Politics and IR, Royal Holloway, University of London. He was previously employed as a Lecturer in International Security at the University of Surrey (2006–12). Tom's research interests lie in European defence and security and British, French, and German defence policy. He is the author of *The Politics of German Defence and Security* (Berghahn, 2007); *Neoclassical Realism and Defence Reform in post-Cold War Europe* (Palgrave, 2010), and *European Defence Cooperation in EU Law and IR Theory* (Palgrave, 2013, with Dr. Theodore Konstadinides).

Ulrike Esther Franke is a doctoral candidate in International Relations at the University of Oxford, supervised by Prof. Sir Hew Strachan. In her dissertation she explores the military implications of the use of unmanned aerial vehicles (UAVs or 'drones') by Western armed forces. Ulrike is a contributing researcher to UN Special Rapporteur on human rights and counter-terrorism Ben Emmerson's research team, working on UAV use in counterterrorism contexts. She has published previously in *Parameters*, *Comparative Strategy*, *Internationale Politik*, *Aus Politik und Zeitgeschichte*, and *RUSI Whitehall Reports*.

Alex Leveringhaus is a political philosopher whose research interests include just war theory and the theoretical foundation of human rights. A graduate of the LSE, he is currently a post-doctoral research fellow in the Oxford Institute for Ethics, Law and Armed Conflict at the University of Oxford, UK, where he works on the 'Military Enhancement: Design for Responsibility and Combat Systems' project. He is also a James Martin Fellow in the Oxford Martin School.

Nicola Power is a Psychology PhD candidate at the University of Liverpool studying Critical Incident Decision Making. Her thesis is looking at the antecedents to decision inertia in strategic decision making settings. She has also been involved in psychological research with the University's remote missile operator simulation suite, developed by the Electrical Engineering and Electronics Department, exploring the cognitive processing of unmanned systems operators.

Jason Ralph is a Professor and Head of the Department of Electrical Engineering and Electronics at the University of Liverpool. He research covers avionics, sensors, and image processing. His interests relate to the utility and reliability of the Human-Machine Interface for remotely-operated air systems.

Regina Rauxloh studied for her first law degree (Erstes Juristisches Staatsexamen) in Germany and received her LL.M. (in Law and Development) and her PhD from the University of Warwick, UK. She has taught at Surrey University international criminal law, public international law, terrorism and law and law of armed conflict, and is now senior lecturer at the University of Southampton. Her research interests lie in International Criminal Law, Corporate Crime, Law of Armed Conflict, and Terrorism. She has published work on corporate criminal responsibility in international criminal law, human trafficking, environmental crimes, comparative procedural law, and regionalism. She has presented her work in New Zealand, the United States, South Korea, Germany, Greece, France, and the UK. Her current work looks at the role which international criminal law should or should not play in a modern global world. This includes the relationship between the ICC and the UN Security Council, especially in relation to military intervention as well as the possible extension of jurisdiction of the ICC.

Tobias Ruettershoff is a PhD Student in Strategy and Security at the University of Exeter under the supervision of Dr Sergio Catignani. His main research interests are in the field of International Relations, with a focus on military and security. In his doctoral thesis, he examines the evolution and use of external expert knowledge in U.S. counterinsurgency campaigns in the last century. Before coming to Exeter, Tobias completed a BA in Political Science and Public Law at the University of Mannheim (Germany) and an MA in International Relations at the University of Sheffield. Tobias is a scholar of the Konrad-Adenauer-Foundation in Germany and has gained work and research experience at the University of Oxford, the Bundeswehr Leadership Centre (ZInFü), and the German Parliament.

Jamie Shea is NATO Deputy Assistant Secretary General for Emerging Security Challenges. He has been working with NATO since 1980. Positions included Director of Policy Planning in the Private Office of the Secretary General, Deputy Assistant Secretary General for External Relations, Public Diplomacy Division, Director of Information and Press, Spokesman of NATO and Deputy Director of Information and Press, Deputy Head and Senior Planning Officer at the Policy Planning and Multilateral Affairs Section of the Political Directorate, as well as Assistant to the Secretary General of NATO for Special Projects. Jamie acts as Professor of the Collège d'Europe, Bruges, Visiting Lecturer in the Practice of Diplomacy, University of Sussex, Associate Professor of International Relations at the American University, Washington D.C. and lectures at the Brussels School of International Studies at the University of Kent. He holds a D.Phil. in Modern History from Oxford University. Amongst his many associations and memberships, Jamie Shea is Member of the Advisory Board, Security and Defence Programmes at

Chatham House, Member of the Policy Council at the World Economic Forum in Geneva and Founder and Member of the Board, Security and Defence Agenda Brussels.

Conway Waddington is a doctoral candidate at the University of Johannesburg, examining moral philosophies relating to war in the context of increasing resource-related conflict. Conway has an MA from the University of KwaZulu-Natal in political philosophy and ethics. He has lectured at UKZN, and presented papers relating to strategic security and military ethics at several international conferences. Conway is a senior analyst at Consultancy Africa Intelligence, and prepares tailored research relating to African regional security, terrorism, and nuclear non-proliferation.

Ina Wiesner is Research Fellow at the Bundeswehr Academy for Information and Communication in Berlin, Germany and a lecturer for military studies and defence policy at the Humboldt University at Berlin and Potsdam University. She received her PhD in Political and Social Sciences from the European University Institute. She recently published *Importing the American Way of War? Network-Centric Warfare in the UK and Germany* (2013) and is the editor of the book *German Defence Politics* (2013). Her research interests include the sociology of military technology, German military transformation, and methodology in the social sciences.

Chris Woods is a London-based investigative journalist specialising in conflict and national security issues. For many years a senior producer with the BBC's Panorama and Newsnight programmes, in 2010 he established and led the Bureau of Investigative Journalism's award-winning drones team. His reports into aspects of the covert use of RPAs have appeared in the *Sunday Times*, the *Guardian*, and the *Independent*, and he was awarded the Martha Gellhorn Prize for Journalism for his work in 2013. His book *Sudden Justice*, which examines the role of armed drones across the War on Terror's many battlespaces, is published by Hurst.

Preface

Precision strike capabilities: political and strategic consequences

Jamie Shea[1]

For the past two decades, the focus of international security policy has been on military expeditionary operations. The world's major security organisations (UN, NATO, EU) have had to grapple with the intractable challenge of trying to make military power, and the billions of dollars needed to deploy it in volume, count in situations where there is not a conventional adversary to defeat, but a range of amorphous actors. Military force has been necessary to establish a foothold and change a regime; but it has not by itself ensured stabilisation and reconstruction. Instead we have needed to develop an integrated approach where the military is able – both materially and conceptually – to fight and rebuild at the same time, as well as support the international and local actors engaged in the civilian arena.

These lessons have been hard-learned, particularly when it comes to the costs in lives, and wounded servicemen and women; not to speak of the US$806 billion spent in Iraq and the (thus far) over US$444 billion spent in Afghanistan by the United States alone (Congressional Research Service, 2011). Moreover, if the intervening powers have learned painfully to adjust their strategy to local conditions and to integrate their military and civilian instruments better, this alone has not been a guarantee of success. In no circumstances has a military force withdrawn from a former conflict zone with a sense of the job being truly done, and a relapse into violence inconceivable. Interventions, and Afghanistan is but the latest example, are a contest between two factors: first, what a military force on a steep learning curve can achieve with the time and resources it is given; second, what the patience of the local population as well as political and public opinion in the intervening countries are prepared to accept. As the timeframe of the latter is inevitably shorter than that of the former, it is hardly surprising that most interventions have ended prematurely (see, for example, Traub, 2012).

The prevailing mood of disillusionment with military interventions has focussed today's security debates on two core concerns. One is avoiding interventions in future by predicting crises better in advance; and heading off eventual conflict through better use of political, economic, and diplomatic instruments. Prevention sounds both attractive and self-evident, but our historical record in anticipating crises and in picking up let alone acting on the early warning signals is a particularly miserable one.

The second concern is that when interventions prove nonetheless inevitable, they should be conducted without incurring most of the usual risks associated with interventions. In other words, without the need to painstakingly assemble an international coalition, without boots on the ground, and without a long-term commitment to stabilisation or nation-building. The advent of modern technology, and particularly pilotless aircraft or drones, as well as precision-guided munitions and missiles, together with highly accurate reconnaissance and surveillance systems, is transforming the battlefield. These technologies allow adversaries to be killed without close-in combat and the attendant risk of casualties on one's own side. They offer states a range of options for limited and containable military actions short of 'all or nothing' scenarios. There is often no need to seek parliamentary approval or to rally political and public support for messy and prolonged ground operations. Precision strike enhances the possibilities of nations being able to act alone, often in covert operations, without needing to declare formal hostilities, or to seek the UN mandates which ground operations, by their very size and visibility, more naturally require.

Drones, in particular, can avoid the perils of escalation once a force is deployed. No exit strategy or mission objectives need to be established. If a drone is shot down or crashes, no pilot is killed or taken prisoner; so no countervailing pressure can be applied by the adversary who has been targeted. Moreover, drones do not require military bases on the territory where the intervention is taking place. Their footprint is minimal, and should the need arise they can be relocated almost overnight from one location to another almost anywhere. Because they can stay over a target area for several hours (unlike a manned aircraft that often can stay for only a very short period and requires expensive air-to-air refuelling), drones can also generate large amounts of intelligence data and over time establish patterns of human behaviour and activity in the area they are monitoring.

Even if no missiles are fired, the mere presence of drones buzzing overhead can keep insurgents or terrorists on the run, and thus less capable of organising attacks. Moreover, the missiles fired from drones can achieve a degree of accuracy that even satellite or laser guided munitions fired from aircraft would have difficulty in achieving. Indeed, the intelligence provided from reconnaissance drones such as Global Hawk can considerably enhance the usability and accuracy of more conventional precise strike munitions fired from aircraft and ships. This, combined with their comprehensive up-to-date intelligence picture, makes a combat drone like a Reaper or a Predator a potentially attractive weapons system for decision makers who are keen to avoid 'collateral damage' and being sucked into costly, protracted, and highly unpredictable ground operations. So, as budget crunches hit home and the costs of interventions in places like Iraq or Afghanistan spiral ever upwards – Linda Bilmes and Joseph Stiglitz of Harvard University have estimated that the US has spent nearly $3 trillion in its various responses to the 9/11 attacks (Bilmes and Stiglitz, 2008; see also BBC News, 2011) – unsurprisingly the use of drones has multiplied. They have appeared also, beyond Afghanistan and Pakistan, in other theatres, such as Yemen, Somalia, and perhaps in the future in the Sahel as well. Press reports indicate that the US has also set up a drone base in Niger (Whitlock, 2013).

From a military vantage point, drones have not fundamentally changed the law of armed conflict and should be considered like any other weapon – subject to the same criteria of last resort, discrimination, and proportionality as all other instruments of killing. The former US Director of National Intelligence, Admiral Dennis Blair, has even described them as 'a long range sniper' (Council on Foreign Relations, 2013). Others have pointed out that their effect is the same as that of firing a missile or dropping a bomb. In such a context, drones even have certain advantages: greater accuracy, more localised targeting, and less 'collateral damage'. Like the guillotine, which was invented for humanitarian motives, they may appear sinister at first but, upon reflection, if a weapon is going to be used, is it not better to use that which is most reliable and the least costly?[2] Weapons reflect the strategic environment in which they are invented as much as they, in turn, determine that environment. Thus drones embody a desire to keep conflicts at a distance, wage them at an acceptable human and financial cost, and deprive enemies of the benefits of being able to operate in small groups and hide in inhospitable areas which conventional armed forces would find it difficult to get to.

As it is not only the United States which is facing these challenges, it is hardly surprising that a number of countries are trying to acquire a similar capability, either by buying from the United States or Israel, or developing their own remote controlled, pilotless aircraft. Recent examples are Turkey, which has requested US drone technology to help it counter its Kurdish insurgency, or France, which has sought to purchase two Reapers from General Atomics for use in the Sahel (Whitlock, 2011; Eshel, 2013). According to an unclassified July 2012 report by the US Government Accounting Office (GAO), 76 countries already have acquired unmanned aerial vehicles. The report also states that 'countries of concern' are attempting to acquire advanced UAVs from foreign suppliers or seeking illegal access to US technology. At least 50 countries are trying to build 900 different types of drones, the GAO claims (GAO, 2012). Given the critical dependency of modern armies on intelligence, reconnaissance, and surveillance assets prior to and during an operation, we can expect that unarmed drones as a cheaper, more versatile, alternative to satellites and manned ground surveillance aircraft (like the US Joint Surveillance

Target Attack Radar System) will be a feature of the inventory of nearly all major armies within a few years. Drones will also be used for domestic security as well; for police forces or in monitoring long insecure borders such as that between the US and Mexico. They will even feature in traffic control and environmental mapping. Yet one word of caution is in order: as often in the past, the introduction of a new technology does not make old ones obsolete. Recently US commanders in the Pacific have taken to resuming U2 flights (a 1950s' technology) because these aircraft operating at 75,000 feet altitude can navigate around cloud more effectively than Global Hawk drones (Roston, 2013).

Currently, drones are the preserve of only a few NATO Allies, and have remained resolutely under national command and control. As such, they have contributed indirectly to NATO's ISAF mission in Afghanistan rather than being a formal component of NATO's order of battle. Drones have, therefore, not been a national capability assigned to NATO, and to be operated by the SACEUR or the NATO multi-national chain of command, as has for example been the case with the deployment of Patriot air defence missiles to Turkey by the US, Germany, and the Netherlands during the conflict in Syria. NATO currently has no policy on armed drones and there are no calls for it to have one. The Alliance has been interested in drones more in the area of Intelligence, Surveillance, and Reconnaissance (ISR). NATO is currently procuring five Global Hawks to have its own dedicated ISR capability as this proved to be a major shortfall during its recent Libya operation, when similar US capabilities were only available intermittently. The question therefore is whether offensive drones will become legitimised through collective ownership and rules of engagement, or be more like tactical nuclear weapons which are nationally owned and operated but where Allies can at least have a say in their tasking and use on the condition of sharing some of the operating and infrastructure costs.

The mounting evidence of drone use has not – thus far at least – provoked the focussed protests or mass demonstrations in the West associated with other innovative weapons in the past: first and foremost nuclear weapons. The latter, of course, represent a totally different order of destructiveness. Yet within the community of strategic analysts, and beyond in international legal circles, parliaments, and the NGO environment, questions are being asked and misgivings are being expressed. The criticisms may be small in scale but they are becoming more vocal. As has often been the case in the past with emerging military technologies, the use of drones has run far ahead of a clear understanding of their long-term political, strategic, legal, and indeed moral consequences. It is this debate which is now starting.

In the first place, there is a concern about secrecy. Although military deployments such as Iraq or Afghanistan may have been difficult and increasingly unpopular, governments have had to announce and justify their strategies; in many cases, parliaments (such as the German Bundestag) had to give their approval to a national participation in the missions; host and transit countries needed to be consulted, rules of engagement clarified and casualties announced. When airstrikes went astray, causing civilian deaths, investigations were launched and compensation paid to the families of the victims. Conflict was at least reasonably transparent, press and media reporting abundant, and those wanting to know what was going on could access multiple sources (including soldiers' e-mail or Facebook messages home) in order to find out. Drones, on the other hand, are still shrouded in a high level of secrecy. Very little is said about the policy behind their use, the rules of engagement or the results, especially in terms of disclosing information on the casualties, whether the immediate targets are reliably identified terrorists or non-combatants accidentally caught up in the strikes. Another issue concerns the so-called 'signature strikes' where individuals or groups with no clear identity are targeted because they behave in suspicious ways or appear to be associated with identified terrorist suspects. If governments do not provide official figures we have to fall back on the estimates of NGOs and research institutes, such as the Long War Journal, the Bureau of Investigative Journalism, and the New America Foundation. The latter reported recently that the civilian casualty rate from drone strikes in Pakistan was 10 per cent in 2012, compared to as high as 60 per cent in 2006 (New America Foundation, 2013). The question to be answered is: how much of this activity can be legitimately secret, and how much can and should be debated within government, and between government and parliaments and broader public opinion?

The second issue concerns what is targeted, and on what basis? Those who have declared war on the United States, for instance, and who are clearly planning and preparing to attack it will find their way eventually onto a target list. But at what point does one decide to attack them? Only when it has proven impossible to capture them and bring them to justice? Or when intelligence gathering indicates that they are about to carry out a terrorist action? Or simply because a clear targeting opportunity presents itself? And what about the criteria for launching a drone strike when these insurgents and terrorists frequently hide among the local population, thereby increasing the risk of civilian casualties, and even if some of these are family members of the targeted individuals? How much obligation should there be to disclose details of possible civilian casualties and to compensate their families as happens frequently after airstrikes?

The current legal argument for drone strikes usually stresses the right of self-defence; but how far does this embrace pre-emption of threats before they have assumed a concrete, imminent form? These are no doubt complicated questions, but it would be useful to have a broader public understanding of the criteria guiding decision making and with a view to reassuring the public that drone strikes are used only when absolutely necessary and against targets defined narrowly rather than broadly. It is key that the easy availability of drone technology should not lead to an unduly broad and permissive definition of who is an 'enemy' and thus a potential target. We should not do things simply because we can do them.

A fourth debate concerns authority. If drones are military weapons used in operations that are defined as war and follow the laws of armed conflict, should they be used in the first instance by military authorities and not by intelligence agencies operating in the guise of covert operations? This connects with a related issue which is whether the role of intelligence agencies should be limited to the collection of intelligence rather than the conduct of military-style operations. At which level of government should the final decision to launch a strike be taken? What kind of prior clearances need to be obtained within government (legal departments as well as intelligence services, diplomatic as well as military inputs) before a strike is deemed 'mature'? From what we currently know, are these clearances sufficiently stringent and intrusive, and is the circle of those who need to be involved narrow enough to ensure confidentiality but broad enough to ensure due process and legitimacy?

A final set of issues brings us to the political domain. No one doubts the technological prowess of drones or their usefulness in fighting the campaign against international terrorism. But the wider political context must also be considered. How useful are drones as an element in winning ultimately the war against terrorism? Do they take us forwards or backwards? Commentators have pointed to the fact that drone strikes can further alienate the local population from the intervening power (see, for example, Mothana, 2013; Zenko, 2013). When launched from abroad and without the formal sanction of the governments over whose territory they are operating, they can appear as a violation of national sovereignty. In this respect, how much should local governments be involved in co-authorising drone strikes? For instance, President Mansour Hadi of Yemen has stated that he personally signs off on every strike by US drones in his country (Al-Shamahi, 2013). Moreover, the strikes may motivate more people to join the ranks of the terrorists out of a desire for revenge. So do they reduce or increase the pool of available recruits? Some experts believe that most terrorists are recruited by the job opportunity or the search for group identity rather than by outrage provoked by drone strikes (Boyle, 2013: 1–29). Yet we also know from previous experience that when terrorist or insurgent leaders are killed, there is always a number two to take their place (Krishnan, 2013: 89–97). This is clearly not a reason not to move against identified and proven terrorists but simply not to expect drone strikes by themselves to eliminate the terrorist threat. They are at best a means of keeping that threat under control or at bay in order to give time for a more comprehensive and long-term counter-terrorism strategy to take effect.[3]

The danger here is that instead of being seen as just one element in that overall strategy, drones become a substitute for it.[4] Precision strike can create a climate where we focus only on the tactical and the immediate and lose sight of our long-term interests and strategic objectives. It may foster the illusion that we can defend against terrorism or other forms of extremism at minimal cost and without the need for direct engagement. It is not particularly original to say that the elimination of extremism requires education, development, jobs, and democratic institution building, but it is nonetheless true. To nations tired of conflicts and nation-building abroad, drones may be initially a welcome relief; but they are not a panacea and they must not drive strategy or options in the way that you may do when you have a hammer, and all problems begin to look like nails. In this respect, it would be interesting to learn not just how many terrorists have been killed but also how effective drones really are in stopping attacks or changing terrorist behaviour over time. Sooner or later we will have to return to the intractable but unavoidable question of how we build local capacity, democratic institutions, and economic development in the places of the world where extremism is most likely to thrive, and whether our foreign and development policies are really adequate to this task.

As these debates evolve, some initial responses are being given. In a speech to the Woodrow Wilson Center in Washington on 30 April 2012, John Brennan, then President Obama's Counter-Terrorism Adviser and now CIA Director, laid out the rationale for drone policy in a way not done before. He stated that drone strikes are legal both under the Authorization for Use of Military Force passed by Congress after the 11 September attacks, and because

> there is nothing in international law that bans the use of remotely piloted aircraft for this purpose or that prohibits us from using lethal force against our enemies outside of an active battlefield, at least when the country involved consents or is unable or unwilling to take action against the threat.
>
> (Brennan, 2012)

More open engagement by government is a good start; but explaining how the US applies this law to drones, rather than simply asserting that the law applies, would set a good example.[5] Brennan has also spoken of issuing a code of conduct for drone strikes that would spell out the criteria for their use. Listing the core criteria informing a political decision to launch a strike would be a step forward. Yet the real reason to have this debate now is to set some basic rules of the road before everyone starts to acquire drones and can use them almost anywhere. As the technology becomes cheaper and more commercially available, drones will be like cyber. Anyone will be able to join the game. Drones will be harder to trace. We must avoid a world where people can be routinely and anonymously targeted. In this respect, there is already in the US a DIY Drones organisation with 45,000 members all engaged in building these systems as a hobby and for private use.[6]

So before the proliferation of drones goes beyond the point of no return, we need to think about how we control and regulate access to drone technology and know-how. Is there a case here for export controls or even a Drone Technology Control Régime as we have for short-range missiles?

At the moment that the 'war of the shadows' is being put under the spotlight and its strategic implications are being debated both among experts and with the wider public, it is useful to have a guidebook to inform us further and to set out clearly and eloquently the key issues that should concern us. This book does not, and cannot attempt to answer all the questions raised in this preface. But it is undoubtedly one of the most comprehensive and detailed to emerge on this topic to date. Following an initial workshop organized by Professor Sir Michael Aaronson, Dr. Wali Aslam, Dr. Tom Dyson, and Dr. Regina Rauxloh at the University of Surrey in 2012, the editors have brought together a number of emerging scholars from disciplines as diverse as international relations, law, philosophy, political science, psychology, and sociology, each with his or her own valuable perspective on the impact of the new technologies on conflict. They present a cross-cutting analysis that does full justice to such a complex issue and deserves the broadest readership among experts and the interested public.

Notes

1 The views expressed in this preface are entirely those of the author alone. They should not be construed as representing an official position of NATO, but are contributed in a purely personal capacity.
2 On the cost of drones vs. manned aerial systems, see 'The US and Its UAVs: A Cost-Benefit Analysis', http://americansecurityproject.org/blog/2012/the-us-and-its-uavs-a-cost-benefit-analysis (accessed 4 November 2013).
3 On the need for less reliance on tactical strikes and greater coherence in strategy, see Nasr (2003).
4 Ibid.
5 On the application of the Authorization for Use of Military Force to drone strikes, see, for example, Aslam (2011: 313–29).
6 See http://diydrones.com/profiles/blogs/diy-drones-at-45–000-members, 2 November 2013 (accessed 4 November 2013).

References

Al-Shamahi, A. (2013) *BBC News* (30 April 2013) 'US Drones Strain on Yemenis' Dual Loyalties', www.bbc.co.uk/news/world-middle-east-22340837 (accessed 1 July 2013).

Aslam, W. (2011) 'A Critical Examination of American Drone Strikes in Pakistan: Legality, Legitimacy and Prudence', *Critical Studies on Terrorism*, 4, 3: 313–29.

BBC News (2011) 'Iraq War in Figures' (14 December 2011), www.bbc.co.uk/news/world-middle-east-11107739 (accessed 1 July 2013).

Bilmes, L. and Stiglitz, J. (2008) *Washington Post* (9 March 2008) 'The Iraq War will Cost Us $3 trillion, and Much More', www.washingtonpost.com/wp-dyn/content/article/2008/03/07/AR2008030702846.html (accessed 13 January 2014).

Boyle, M. (2013) 'The Costs and Consequences of Drone Warfare', *International Affairs*, 89, 1: 1–29.

Brennan, J. (2012) Wilson Center (30 April) 'The Ethics and Effectiveness of US Counterterrorism Strategy', www.wilsoncenter.org/event/the-efficacy-and-ethics-us-counterterrorism-strategy (accessed 1 July 2013).

Congressional Research Service (29 March 2011) 'The Cost of Iraq, Afghanistan, and Other Global War on Terror Operations since 9/11', www.fas.org/sgp/crs/natsec/RL33110.pdf (accessed 1 July 2013).

Council on Foreign Relations (22 January 2013) 'US Drone Strike Policies', www.cfr.org/counterterrorism/us-drone-strike-policies/p29849 (accessed 4 November 2013).

Eshel, T. (2013) *Defense Update* (31 May) 'France Opts for US Reapers', http://defense-update.com/20130531_france_opts_for_the_reaper.html (accessed 5 January 2014).

GAO (US Government Accounting Office) (July 2012) 'Non-proliferation: Agencies Could Improve Information Sharing and End-Use Monitoring on Unmanned Aerial Vehicle Exports', www.gao.gov/assets/600/593131.pdf (accessed 4 November 2013).

Krishnan, A. (2013) 'Dead on Target? The Strategic Dead End of Targeted Killing as a Way of War', in Aaronson, M. and Johnson, A. (eds) 'Hitting the Target: How New Capabilities are Shaping International Intervention', *RUSI Whitehall Report*, 2: 89–97.

Mothana, I. (2013) *New York Times* (13 June 2013) 'How Drones Help Al-Qaeda', www.nytimes.com/2012/06/14/opinion/how-drones-help-al-qaeda.html?_r=0 (accessed 21 November 2013).

Nasr, V. (2013) *The Dispensable Nation: American Foreign Policy in Retreat*, New York: Doubleday.

New America Foundation (2013) 'Analysis: The Drone War in Pakistan', http://natsec.newamerica.net/drones/Pakistan/analysis (accessed 1 July 2013).

Roston, A. (2013) *Defense News* (24 June 2013) 'Will Congress Let the Air Force Abandon the Global Hawk?', www.defensenews.com/article/20130624/C4ISR01/307010011 (accessed 4 November 2013).

Traub (2012) *New York Times* (18 February 2012) 'The End of American Intervention', www.nytimes.com/2012/02/19/opinion/sunday/the-end-of-american-intervention.html?pagewanted=all&_r=0 (accessed 1 July 2013).

Whitlock, C. (2011) *Washington Post* (11 September 2011) 'US Considering Ankara's Request to Base Predators in Turkey to Fight a Kurdish Group in Northern Iraq', www.washingtonpost.com/world/national-security/us-considering-ankaras-request-to-base-predators-in-turkey-to-fight-a-kurdish-group-in-northern-iraq/2011/09/09/gIQA5R1fIK_story.html (accessed 1 July 2013).

Whitlock, C. (2013) *Washington Post* (22 March 2013) 'Drone Base in Niger Gives US a Strategic Foothold in West Africa', www.washingtonpost.com/world/national-security/drone-base-in-niger-gives-us-a-strategic-foothold-in-west-africa/2013/03/21/700ee8d0–9170–11e2–9c4d-798c073d7ec8_story.html (accessed 4 November 2013).

Zenko, M. (2013) Council on Foreign Relations (January) 'Reforming US Drone Strike Policies', www.cfr.org/wars-and-warfare/reforming-us-drone-strike-policies/p29736 (accessed 21 November 2013).

Acknowledgements

The authors are very grateful to the University of Surrey's Institute of Advanced Studies who funded the July 2012 workshop ('Hitting the Target: How New Capabilities are Shaping International Intervention') where the foundations of this volume were set in place. They would like to thank, in particular, the IAS coordinator Mirela Dumic, for her valuable help with arranging the workshop. The authors would also like to thank the University of Surrey's Centre for International Intervention (cii) for its continuing support and Katharine Wright, PhD candidate in the Politics Department at the University of Surrey, for her compilation of the index. Finally, the authors would like to express their gratitude to the editorial staff at Routledge, the Routledge Global Security Studies series editors, Professors Aaron Karp and Regina Karp, as well as the anonymous reviewers whose comments were of great help in improving the book.

Part I

Strategic and foreign policy drivers and consequences

Introduction

Precision strike warfare and international intervention: strategic, ethico-legal, and decisional implications

Mike Aaronson and Tom Dyson

Since the end of the Cold War US technological superiority has led to a more proactive and, some would argue, aggressive, approach to international military intervention. New technologies, including the capacity to mount precision military strikes from high-level bombing campaigns – and more recently the selective targeting of individuals from armed unmanned aerial vehicles (UAVs) – have facilitated air campaigns, supported by Special Operations Forces, without the commitment of large numbers of troops on the ground. Such campaigns include NATO's high-level aerial bombardment of Milošević's forces in Kosovo in 1999 and of Gaddafi's in Libya in 2011. The development of armed UAVs and electronic data intercept technologies has further expanded the potential scope of interventions against Islamic militants in the Tribal Areas of Pakistan, Somalia, and Yemen. In addition, the US has undertaken targeted manned operations involving Special Operations Forces against selected targets such as Osama Bin Laden.

The aim of this book is to explore whether – and if so how – new capabilities made possible by precision strike technologies are reshaping approaches to international intervention. It will examine three key and interrelated dimensions of precision strike capabilities: the strategic and foreign policy drivers and consequences; the ethical and legal implications of new capabilities, and, finally, the implications for decision making at the strategic, operational, and tactical levels.

Precision strike warfare, the revolution in military affairs, and post-Cold War interventionism

'Precision strike' technology has been around for some considerable time. It is closely associated with the 'Revolution in Military Affairs' (RMA). The RMA emphasizes the emergence, at key periods throughout history, of revolutionary advances in technology and/or military organisation which can deliver a decisive advantage for militaries capable of harnessing them. The roots of the RMA lie in the Vietnam War when the US deployed, for the first time, battlefield sensors, satellite navigation, and radiation-seeking munitions, and began to develop superiority over its competitors in the electronic battlefield (Cohen, 2004: 396, 399; Morgan, 2000: 133). The US military and policy makers were committed to avoiding a repeat of the attrition of the Vietnam War and the technologies associated with this emerging RMA offered the apparent opportunity to deliver the capacity to achieve decisive victory over opponents (Stone, 2004: 417).

The lessons drawn by the US military and policy makers from the most prominent conflict of the post-Cold War era – the 1991 Gulf War – also support these conclusions. US victory in the Gulf War was heavily dependent upon the deployment of ground forces. However, the manner in which airpower had been used to target critical infrastructure of the Iraqi regime appeared to demonstrate the potential of stand-off[1] precision strike air power and satellite-guided, automated command, control, and communications (C3) to achieve operational goals with greater efficiency (Bratton, 2002: 88; Morgan, 2000: 133; Mahnken and Fitzsimmonds, 2003: 121).

Hence following the 1991 Gulf War the technologies associated with RMA were championed by a number of strategic planners within the Pentagon and US military. The Air Force and Navy were particularly vociferous proponents of the advantages of RMA technologies and doctrine, in no small part due to the enhancement in defence budget share that the RMA would offer these services (Bratton, 2002: 88). Furthermore, the RMA was particularly attractive to policy makers due to the apparent advantages it offered to tackle 'rogue states' which were identified as the key opponents to the US in the May 1997 Quadrennial Defence Review (Metz, 2006: 5). For Presidents Bill Clinton (1993–2001) and George Bush (2001–09) the RMA presented the opportunity to facilitate the use of military force in the context of a high-level of public sensitivity to troop losses and civilian casualties (Stone, 2004: 418). Crucially, however, the RMA was viewed as a vital strategic imperative due to the advantages it appeared to deliver in enhancing the speed, agility, and manoeuverability of the US military. US policy makers feared that the technological advances of the RMA would be harnessed by competitors to the US, should it fail to remain at the forefront of the RMA (Kagan, 2006: 217). In short, the RMA was perceived by US policy makers as critical in ensuring the US could cement its position as the unipolar leader in the post-Cold War era.

The RMA was, therefore, central in fostering the optimism that underpinned a spirit of liberal interventionism that prevailed among the US and its allies in the post-Cold war era (Sabaratnam, 2011: 13–31). This period had its high (and some would say, low) point with the 1999 NATO campaign to remove Slobodan Milošević from Kosovo, effected principally by means of high altitude bombing using 'precision strike weapons'. That this intervention was conducted in the absence of a UN Security Council resolution – a requirement stipulated in the UN Charter – was one element in the discrediting of the concept of 'humanitarian intervention' and the emergence in 2001 of the more nuanced principle of the 'Responsibility to Protect'. However, R2P, as it is known, has yet to acquire the same justificatory power with regards to intervention as occurred during the 1990s in places such as Northern Iraq, Somalia, Bosnia, Liberia, Sierra Leone, and Kosovo, to name but the most high-profile cases. During this time, and within the context of 'precision strikes', mention should also be made of the establishment in 1991 and 1992 of no-fly zones over northern and then southern Iraq, and operation Desert Fox in December 1998, aimed at degrading Saddam's capability to manufacture, store, and deploy weapons of mass destruction.

In parallel the world witnessed the emergence and growth of Islamic fundamentalism and transnational terrorism practised by Al-Qaeda and its affiliates, fed by grievances flowing from sources as diverse as the Soviet occupation of Afghanistan, the presence of US troops in the holy territory of Saudi Arabia as part of Operation Desert Storm, and the West's continuing failure to act decisively to provide justice for the Palestinian people. Following the first, unsuccessful, attack on the World Trade Centre in New York in February 1993, Osama bin Laden attacked the US embassies in Kenya and Tanzania in August 1998. The Clinton administration retaliated by bombing the Al-Shifa chemical factory in Khartoum, again with 'precision' weaponry from high altitude (although it has never been clear that the factory was involved in the production of chemical weapons as the Administration claimed). In turn Al-Qaeda mounted a suicide bombing attack on the guided missile destroyer the USS Cole in the Yemeni port of Aden in October 2000. This phase culminated with the devastating attacks of 11 September 2001 (9/11) in New York and Washington D.C. and the subsequent pronouncement by the Bush Administration of the 'Global War on Terror'.

Following 9/11 the US invaded Afghanistan to remove the Taliban regime that had sheltered and subsequently refused to surrender bin Laden. Initially waging a proxy war by supporting the Northern Alliance, the US and allies subsequently found themselves committed to a major expeditionary campaign, fighting an insurgency as well as conducting anti-terrorist operations. In parallel – and with serious consequences for the Afghan campaign – in 2003 the Bush Administration invaded Iraq on the grounds (which proved to be unfounded) that Saddam Hussein continued to possess weapons of mass destruction. Both the Iraq and the Afghanistan campaigns became major millstones for the US, consuming vast amounts of time, energy, and money and costing significant numbers of American and allied lives, as well as many more among the local populations of the two countries. Both also laid bare the limitations of the RMA and its exaggerated faith in the power of technology and superior manoeuvrability to permit the deployment of many fewer boots on the ground (Betz, 2007: 238).

Despite the limitations of precision strike capabilities revealed in Iraq and Afghanistan, precision strike technologies associated with the RMA remain central to US security policy under President Barack Obama. As a Senator, Obama was fiercely critical of the Iraq campaign, which he labelled a 'dumb war' (Obama. 2002). Although as President he initially appeared to believe that the Afghanistan campaign could be won – authorising in 2010 a significant 'surge' in the numbers of US forces in theatre – eventually he became much more wary about the wisdom of the large-scale expeditionary warfare that both these campaigns entailed. However, even before he assumed office, he had been adamant about his resolve to 'take out the terrorists', wherever they threatened US security (Obama. 2007). The increasing refinement of remote weaponry as a means of achieving this goal supported this policy ambition. Thus, under Obama, while large-scale expeditionary operations have become much less likely, the use of targeted killings using UAVs as an instrument of US security policy has increased dramatically, including in countries where the US is not otherwise engaged in armed conflict, for example in Pakistan, Yemen, and Somalia.

However, there is a strong body of opinion that precision strike campaigns are counter-productive because of the resentment they create among the civilian population in the target areas, the so-called 'collateral damage' consisting of civilian casualties and damage to private property (see, for example, Stanford/NYU 2012; Boyle, 2013). While the US has remained adamant that it only attacks military targets, albeit accepting that some civilian casualties do occur, it is clear that the basis on which it makes these claims is itself open to dispute. In other words, there is still much room for debate about the effectiveness, as well as the legitimacy, of these methods.

Questions are also asked about the ethics of the growing use of precision strike weapons, especially UAVs. From the traditional perspective of the right to go to war (*jus ad bellum*), does the existence of the new capabilities affect the calculus of risk versus reward and make more likely a decision to resort to the use of force? In terms of the just conduct of war (*jus in bello*), can these new weapons be used satisfactorily in accordance with the long-standing principles of necessity (use of minimum force levels needed to achieve aims), distinction (discriminating between legitimate targets and innocent civilians), proportionality (collateral damage in proportion to military gain), and humanity? Can deliberate targeted killing outside a formal declaration of war be justified and what level of collateral damage, if any, is tolerable? In terms of the just consequences of war (*jus post bellum*), does this new form of warfare, conducted remotely with a high possibility of collateral damage, make 'winning the peace' harder once the conflict has ended?

It is clear that contemporary international law with regard to armed conflict, perhaps unsurprisingly, is only partly able to provide answers to these questions. While at one level it is very detailed and comprehensive, at another it is very open to differing interpretations.

Thus, both the Bush and the Obama Administrations have argued that the US is involved in a 'transnational global conflict' with 'illegal combatants' and that it is legally entitled to undertake targeted killings of members of these groups wherever they are located. As a leaked 2011 US Department of Justice (DOJ) White Paper advising on the legality of using lethal force against a US citizen suspected of terrorism against the United States argues: 'a lawful killing in self defence is not an assassination'. The DOJ's position is that

> (i) [where] the targeted individual poses an imminent threat of violent attacks against the United States, (ii) capture is infeasible, and the United States continues to monitor whether capture is infeasible, and (iii) the operation would be conducted in a manner consistent with applicable laws of war principles

lethal force would be justified (US Department of Justice, 2013).

Note that in order to defend this position the US has had to adopt a position on 'imminence' of attack that represents a significant departure from traditional Websterian theory and principles ('necessity of self defence, instant, overwhelming, leaving no choice of means and no moment for deliberation'[2]). The DOJ White Paper justifies this with reference to the different nature of the threat represented by Al-Qaeda, which 'demands a broader concept of imminence in judging when a person continually planning terror attacks presents an imminent threat, making the use of force appropriate'. The use of the phrase 'the applicable laws of war' is also open to differing interpretations. Until such time as a case is taken, either to the International Court of Justice as a dispute between states (unlikely), or to the International Criminal Court (ICC) as a prosecution of an individual (even less likely given that the US has not ratified the Rome Statute establishing the ICC and has a veto on the Security Council, which can also ask the Prosecutor to investigate) and a legal verdict reached, these will remain areas where lawyers from different corners will argue different positions with equal passion and conviction.

Under international humanitarian law (IHL) the introduction of new weapons is subject to international law review (Article 36 of Additional Protocol 1 to the Geneva Conventions).[3] The onus is on states to demonstrate that such weapons do not infringe on the core IHL principles of necessity, distinction, proportionality, and humanity. As has been highlighted above, there is nothing technically new about precision strike weapons; rather it is their relatively recent and highly selective use to carry out targeted killings that has given rise to concerns about moral perception and responsibility among those who operate them. This is in contrast, for example, to instruments such as chemical and biological weapons, which are deemed to be inherently inconsistent with IHL principles and are therefore banned by international treaty.[4]

It is worth pointing out that thus far this discussion has assumed a situation of asymmetric warfare, i.e. where a small group of powerful states possesses the new capabilities, but their enemies do not. This raises the for now hypothetical but nevertheless highly important question of whether the judgements they make and the conclusions they reach would be different if they knew these capabilities could be used by others against them. Although this issue is not examined further in this volume, it is in a sense the 'elephant in the room' and will surely be of concern both to academics and to policy makers in the future.

Finally, there are important questions to be addressed about the impact of the new technologies on the human operator and the implications for decision making at strategic, operational, and tactical levels. In addition to the *jus ad bellum* arguments about whether the new technologies lower the threshold for a decision to go to war, there is also a set of issues about how they impact on operational and particularly tactical decisions taken in situations of ongoing armed conflict. Arguments about how remoteness affects operators' decision making abilities are often trivialised with unscientific claims of a 'Playstation mentality'. Whether on the one hand deprivation of sensory cues leads to moral disengagement, or on the other remoteness allows decision inertia to be overcome, encourages rational and analytic processing of data, and removes the potential for human error – these are important questions that merit further study rather than being the subject of uninformed assumptions. It is clear that this is an area with considerable scope for empirical research that would help to determine competing claims made at present on the basis of very little hard evidence.

The strategic, ethico-legal, and decisional implications of precision strike capabilities

This, therefore, is the policy context within which the ethical, legal, and behavioural aspects of 'precision strikes' must be considered, and with which this volume is concerned. The contributors attempt to answer three important sets of questions: What are the strategic and foreign policy drivers and consequences of the use of precision strike capabilities? What legal and ethical implications arise from their use? Finally, what are the implications for decision making at strategic, operational, and tactical levels? The book is accordingly divided into three parts, but many of the issues recur across all three. The authors are from a range of academic disciplines and include early career researchers as well as established experts in their own field.

The first part of the book focuses on the strategic and foreign policy drivers and consequences of precision-strikes both for states undertaking intervention and for those states at the receiving end of precision strikes. As the previous section has highlighted, the US has been at the forefront of developing the precision strike capabilities associated with the RMA. However the short- and particularly longer-term consequences of the use of these capabilities for the US remain unclear. **Andre Barrinha** and **Luis da Vinha** undertake an analysis of the short- and long-term strategic utility of the use of precision strike capabilities by the US and whether what began as a policy-enabler in the War on Terror is now becoming a driver of policy. They employ Ulrich Beck's concept of a world 'risk society' to highlight inconsistencies in the rationale and methods of precision strikes by the US (Beck, 1992). These lead to a spuriously precise assessment of threats, the depoliticisation and micro-management of war, and the effective transfer of risk to the security subjects of the intervened upon countries.

The US, of course, does not enjoy a monopoly on precision strike capabilities. Other states, particularly the European allies of the US in NATO, have also developed a substantial precision strike capability (Terriff *et al.*, 2010). **Tom Dyson**'s chapter focuses on the selective emulation of the RMA by Europe's leading military states (Britain, France, and Germany) and analyses the factors determining the speed and pace of this emulation. Dyson undertakes an examination of the reforms which have taken place to the force structures, military doctrines, and capabilities of these three states. In so doing, he assesses the relative capacity of these states to undertake the precision strike, networked, operations which are associated with the RMA. The extent of European emulation of the RMA is timely, given the post-Cold War growth of the EU as an increasingly autonomous military actor within its geopolitical neighbourhood – a trend that is likely to accelerate as the US shifts its focus away from Europe to the Asia-Pacific region in the context of the rise of China (Biscop, 2012: 1309).

Ulrike Franke deals more specifically with the manifestation of precision strike capability that currently commands greatest public attention: UAVs, or 'drones'. Mapping their dramatic proliferation since the turn of the last century she establishes that between 56 and 78 countries now possess this capability, although only three of these are known users of armed UAVs (the US, Israel, and the UK). She argues that this is surprising given the lack of data proving UAVs' military worth and suggests that cultural factors may be responsible. Drones are apparently becoming military 'must-haves' on the basis that all militaries consider they need unmanned aerial capabilities in order to appear modern. Therefore, despite strongly voiced concerns about how they are used, efforts to ban them or impose significant controls on their use through existing and new arms control instruments are likely to encounter resistance.

The lack of consensus about the political legitimacy of drone strikes has generated fierce public debate about them, both in states undertaking intervention and in states at the receiving end of such strikes. This debate has been highly politicised, with a variety of actors seeking to dominate agenda-setting on developing the public narratives emerging around the issue of drone strikes. Using Constructivist theory, **Wali Aslam** provides insight into the process of the political appropriation of drone-related casualties by political leaders in Pakistan and of the 'war on terror' by politicians in the US and explores the impact of this on national strategy. He demonstrates that in both countries the drone has been used as a means of promoting a particular political agenda and that in the process many other issues have been deliberately obscured.

This leads into the second section of the book, which focuses on the extent to which the use of precision strikes from manned and unmanned systems raises new ethical and legal issues. Continuing with the theme of civilian casualties arising from drone strikes in Pakistan, **Chris Woods** offers a journalist's perspective on the implications for those at the receiving end of these precision strike operations. Drawing on new empirical research conducted in Pakistan by the Bureau of Investigative Journalism, Woods forensically examines conflicting claims about civilian casualties made by both proponents and opponents of drones and demonstrates that these arise not so much from disputes about the numbers killed but rather from differing interpretations of what constitutes a civilian as opposed to a combatant. Coupled with the lack of transparency by the US about these operations this makes it very difficult to make definitive assessments of the merits of US policy. However, Woods suggests that the evidence suggests that collateral damage is not being sufficiently mitigated by the US, and casts serious doubt on the efficacy of armed UAVs in limiting civilian casualties and in achieving the longer-term strategic goals of the US.

Conway Waddington, on the other hand, from the perspective of moral philosophy, argues that quantitative claims about collateral damage should not form the basis of the debate about the ethics and morality of the use of precision strike technology, which – he also reminds us – is not limited to the use of drones. Instead, he considers the process of ethical reasoning that leads to the establishment of a moral position on a policy of targeted killings and the proportionality equations that govern those strikes, suggesting that the very availability of the new capability may be subverting moral restraints on the use of force. He highlights what he claims is the questionable adherence by the US to long-standing norms of proportionality and distinction and argues that the emergence of so-called 'signature strikes' and 'double-tap' tactics are further symptoms of an ongoing moral deterioration within the US strategic security culture.

The contribution of **Tobias Ruettershoff** illustrates that even on the (relatively) safer ground of international law there is still room for extensive debate about the status of the armed conflict in which the US finds itself engaged against Al-Qaeda and its affiliates. Concentrating on the *jus ad bellum* – the right to go to war – he considers whether the US is justified under international law in carrying out drone strikes against a non-state adversary such as Al-Qaeda in Pakistan. He argues that although many such strikes are probably illegal because they do not satisfy the requirements of necessity, immediacy, and proportionality, the interpretation of Article 51 of the UN Charter has evolved in recent years to allow acts of preventive self-defence against non-state actors and that in principle the US is therefore within its rights to carry out such attacks.

Ina Wiesner's chapter seeks to broaden the discussion about the use of UAVs to include their potential for enhancing international security. Pointing out that while proponents of drones emphasise their efficiency and effectiveness – in that they relieve military personnel from a number of combat and support functions – their opponents attack them on grounds of illegitimacy – because they are claimed to violate the laws of war. Taking as a starting point the concept of the Responsibility to Protect she suggests that the capabilities of drones could be used to provide human protection in situations of conflict and potential mass atrocity crimes and that this might encourage powerful states with casualty-averse societies to engage militarily in such missions, from which they fight shy at present.

Finally, in the third part of the book, we consider to what extent the new precision strike technological capabilities impact positively or negatively on the human operator. **Adrian Banks** and **Mandeep Dhami** approach this question from the perspective of experimental psychology. They review models of decision making that have been tested in other contexts and consider how they might be applicable to military decisions that have to be taken at strategic, operational, and tactical levels. They conclude that while the complex and uncertain nature of many of these decisions suits a more intuitive strategy of reasoning, their unfamiliar nature suggests a more analytic approach is needed. Therefore, it is extremely difficult to prescribe one single model; what is appropriate is a flexible mixture of both intuitive and analytic approaches, a mode of thinking known as quasirationality. They suggest that further research is needed into exactly what approaches are taken, and how they might best be supported.

Taking a different theoretical approach, the chapter by **Nicola Power**, **Laurence Alison**, and **Jason Ralph** uses the framework of traditional Just War Theory to address the ethical dimension of the use of precision strike technology while exploring the psychological impact of unmanned weapons systems on military decision making before, during, and after conflict. They suggest that the very novelty of unmanned weapons systems can bias the consideration of ethical issues. They query whether military decision makers have an appropriate level of understanding and expertise in using these systems, importantly when considering the trade-off between short-term tactical advantage and long-term strategic goals, and offer recommendations for further research.

Finally, **Alex Leveringhaus** and **Tjerk de Greef** look ahead to the likely future development of increasingly autonomous targeting systems supporting precision strike capabilities. From the perspectives of moral philosophy and cognitive systems engineering they discuss the ambiguous concept of 'autonomy' and conclude that fully autonomous systems are suitable in some contexts but not others, where on ethical grounds it will always be necessary to have a human operator making key decisions. They advocate an e-Partnership approach to the design of targeting systems, allowing a degree of operational autonomy to the machine while leaving the human in the decision making loop. They argue that, by combining the respective strengths of machines and humans, this approach has significant potential to enhance targeting decisions.

In summary, this edited volume breaks new ground by integrating the perspective of a variety of disciplines – including security studies, international relations, law, philosophy, psychology, and engineering and delivers a thorough examination of the pitfalls and potential benefits of precision strike capabilities for international intervention. These are addressed in detail in the book's concluding chapter by **Wali Aslam** and **Regina Rauxloh**. In keeping with the volume's interdisciplinary approach, Aslam and Rauxloh also explore the key directions for future research on precision strike warfare, both within and between the range of disciplines represented by our contributors.

Notes

1 A 'stand-off' advantage is one that allows the projection of military force without the threat of being targeted by an opponent's defence system.
2 'British-American Diplomacy – The Caroline Case', http://avalon.law.yale.edu/19th_century/br-1842d.asp (accessed 11 March 2013).
3 'Protocol Additional to the Geneva Conventions of 12 August 1949', www.icrc.org/ihl.nsf/FULL/470?OpenDocument (accessed 7 March 2013).
4 'Convention on the Prohibition of the Development, Production, Stockpiling, and Use of Chemical Weapons and their Destruction, Paris 13 January 1993', www.icrc.org/ihl.nsf/FULL/553?OpenDocument (accessed 7 March 2013).

References

Aaronson, M. (2013) 'Interventionism in US Foreign Policy from Bush to Obama', in Bentley, M. and Holland, J. (eds) *Obama's Foreign Policy and the War on Terror*, Abingdon: Routledge.

Beck, U. (1992) *Risk Society: Towards a New Modernity*, London: Sage.

Betz, D. (2006) 'The More You Know, the Less You Understand: The Problem with Information Warfare', *Journal of Strategic Studies*, 29, 3: 505–33.

Betz, D. (2007) 'Redesigning Land Forces for Wars amongst the People', *Contemporary Security Policy*, 28, 2: 221–43.

Biddle, S. (2004) *Military Power: Explaining Victory and Defeat in Modern Battle*, Princeton: Princeton University Press.

Biscop, S. (2012) 'The UK and European Defence: Leaving or Leading?', *International Affairs*, 88, 6: 1297–313.

Boot, M. (2006) *War Made New: Weapons, Warriors and the Making of the Modern World*, New York: Gotham Books.

Boyle (2013) *Guardian* (7 January 2013) 'US Drone Attacks "Counter-Productive", Former Obama Security Adviser Claims', www.theguardian.com/world/2013/jan/07/obama-adviser-criticises-drone-policy (accessed 7 March 2013).

Bratton, P. (2002) 'France and the Revolution in Military Affairs', *Contemporary Security Policy*, 23, 2: 87–112.

Cohen, E. (2004) 'Change and Transformation in Military Affairs', *Journal of Strategic Studies*, 27, 3: 395–407.

Farrell, T. (2008) 'The Dynamics of British Military Transformation', *International Affairs*, 84, 4: 777–807.

Frank, A. (2004) 'Get Real: Transformation and Targeting', *Defence Studies*, 4, 1: 64–86.

Howard, M. (1983) *Clausewitz*, Oxford: Oxford University Press.

Kagan, F. (2006) *Finding the Target: The Transformation of American Military Policy*, New York: Encounter Books.

Mahnken, T. and Fitzsimonds, J. (2003) 'Revolutionary Ambivalence: Understanding Officer Attitudes to Transformation', *International Security*, 28, 2: 112–48.

Metz, S. (2006) 'America's Defence Transformation: A Conceptual and Political History', *Defence Studies*, 6, 1: 1–25.

Morgan, P. (2000) 'The Impact of the Revolution in Military Affairs', *Journal of Strategic Studies*, 23, 1: 132–62.

Obama, B. (2 October 2002) 'Transcript: Obama's Speech against the Iraq War', www.npr.org/templates/story/story.php?storyId=99591469 (accessed 7 March 2013).

Obama, B. (1 August 2007) 'Obama's Speech at Woodrow Wilson Center', www.cfr.org/elections/obamas-speech-woodrow-wilson-center/p13974 (accessed 12 January 2014).

Sabaratnam, M. (2011) 'The Liberal Peace: An Intellectual History of International Conflict Management, 1990–2010', in Campbell, S., Chandler, D., and Sabaratnam, M. (eds) *A Liberal Peace: The Problems and Practicalities of Peacebuilding*, London: Zed Books, pp. 13–31.

Stanford/NYU (2012) International Human Rights and Conflict Resolution Clinic at Stanford Law School and Global Justice Clinic at NYU School of Law: 'Living under Drones: Death, Injury and Trauma to Civilians from US Drone Practises in Pakistan', http://livingunderdrones.org/wp-content/uploads/2012/10/Stanford-NYU-LIVING-UNDER-DRONES.pdf (accessed 7 March 2013).

Stone, J. (2004) 'Politics, Technology and the Revolution in Military Affairs', *Journal of Strategic Studies* 27, 3: 408–27.

Terriff, T., Osinga, F., and Farrell, T. (eds) (2010) *A Transformation Gap? American Innovations and European Military Change*, Stanford: Stanford University Press.

US Department of Defense (2003) *Transformation Planning Guidance* (*TPG*), p. 5.

US Department of Justice (2013) 'Lawfulness of a Lethal Operation Directed against a US Citizen Who Is a Senior Operational Leader of Al-Qa'ida or An Associated Force', http://msnbcmedia.msn.com/i/msnbc/sections/news/020413_DOJ_White_Paper.pdf (accessed 7 March 2013).

1 Dealing with risk

Precision strikes and interventionism in the Obama Administration

Andre Barrinha and Luis da Vinha

Introduction

Since taking office, the Obama Administration has significantly expanded the use and geographical scope of precision strikes, authorising missions ranging from Libya to Pakistan. While very effective in the short term, the use of precision strike capabilities, particularly Unmanned Aerial Vehicles (UAVs), raises doubts as to the long-term consequences of these strikes, especially regarding the social and political effects in the countries where those strikes take place. Drawing on Ulrich Beck's world risk society approach, this chapter highlights how the current US military intervention policy attempts to combine a 'risk management' (Coker, 2009) logic with a precise threat assessment exercise in which the enemy has to be clearly identified and dealt with. This, it will be argued, leads to the micro-management of war in the context of a risk society that depoliticises the use of military force, and transfers the risk associated with its use to the security subjects of the intervened upon countries.

'The weapon of choice' is an expression commonly associated with the usage of UAVs by the first Obama Administration and which accurately portrays how the use of precision strikes has moved up the ladder of US strategic priorities (see Brennan, 2012; Gates, 2010). UAVs are no longer mere complementary pieces of a larger military puzzle; they are its core: the starting point from which everything else is enabled.

Technological developments help to explain why President Obama has decided to invest intensively in the use of UAVs. This is, however, insufficient to enable us to understand how they are used. Technology alone fails to explain why precision strikes have been prominently transformed into targeted killings or why Washington uses these new weapons against targets in countries that are not at war with the US.

As argued in this chapter, there is a common rationale to why Washington is using UAVs for targeted killings that is better understood if one resorts to the risk literature, namely to the concepts of risk society (Beck, 1992, 1998), risk wars (Beck, 2009; Coker, 2009; Rasmussen, 2006) and, within the latter, risk-transfer (Shaw, 2005).

Developed by authors such as Ulrich Beck, Anthony Giddens, and Niklas Luhmann, the idea that we live in a risk society has taken centre stage in Social Sciences, most significantly in Sociology, the disciplinary home of these authors, but also in International Relations and Security Studies. Its core premise states that 'we increasingly live on a high technological frontier which absolutely no one completely understands and which generates a diversity of possible futures' (Giddens, 1998: 25). In other words, we live in a society in which our actions have unexpected consequences and risks that are magnified by technological developments. Consequences that we unsuccessfully attempt to manage end up generating further risks: 'There is [...] no end to risks. For every attempt to remove a risk, new risks will proliferate' (Rasmussen, 2001: 292). These risks are at once unpredictable and manufactured (Beck, 2009). They are the unknown unknowns of our own making.

Risks are also marked by their inexistence. A risk that is materialised is no longer a risk, but an accident, tragedy, or catastrophe. In that sense, it is always a future event, something that may (or may not) occur (Beck, 2009: 9). Accordingly, risks reframe how societies deal with time. We deal with the present while trying to control the future (Rasmussen, 2001: 293). We live in reflexive late modern societies that are in constant need of predicting and safeguarding the future in the present time. Therefore, we live in societies that have moved beyond the linear faith in progress and have turned inwards, constantly assessing, predicting, and controlling their own existence (collectively and individually). They are anxious societies that are in constant fear of themselves.

In that regard, the wars these societies wage are also risk wars. As addressed in the following section, risk wars are 'risk management [activities] in all but name' (Coker, 2009: viii). They combine different forms of violence (ibid.: 147) and generate different responses when compared to modern warfare. They address future events (the ethnic cleansing that is to occur, the terrorist attack that is to take place, or the nuclear weapon that is to be built) and they are fought in ways that remove the direct risks of hostilities from the soldiers waging it. Thus, both the *jus ad bellum* and the *jus in bello* are intimately linked with the symptoms of the risk society: the anxiety about the future and necessary will to control the present, in this case to control the potentially unpredictable consequences of war leads society to resort to precision strikes. These elements of risk society are, in our view, clearly present in Obama's policy with regard to precision strikes.

Despite the high expectations of change, the Obama Administration did not radically alter the previous administration's general counterterrorism policy; in a sense, it has actually radicalised it (Indyk *et al.*, 2012; Lindsay, 2011). Even if modern practices of invasion and use of 'troops on the ground' visible in Afghanistan and Iraq have been drastically reduced by Obama, the corresponding precision strike practices remained in place, were geographically extended to other countries, and were intensified where they were already employed. Though officially over, the 'Global War on Terror' has become a risk war in its totality. In that regard, Obama has used the same political technique his predecessor used in 2003 when a month after the invasion of Iraq he declared the war to be over (Coker, 2009: 9). It is the symbolic politics of risk society (Beck, 1992), which consists of the 'art of producing closure where there is none' (Rasmussen, 2006: 37).

Framing the Obama Administration's precision strike policy within this particular social understanding of the world makes it possible to highlight some of the contradictions that are present in this policy, particularly the temptation to micro-manage drone strikes through the definition and implementation of target lists. As argued here, regardless of the questionable legal issues associated with this policy (Kessler and Werner, 2008), from a risk perspective, micro-management is an attempt to control the future by transforming risks into individualised threats, which subsequently generate new risks.

In terms of structure, this chapter starts by delving into the concept of risk society and its penetration into the field of Security Studies. It assesses how war is treated from such a perspective and how international terrorism – i.e. the main risk against which Obama has used precision weapons, particularly drones – acquires a particular importance in that regard. Section two will focus precisely on the recent evolution of US precision strike policy, highlighting the articulation between target lists, drones, and terrorism. This will pave the ground for a more in-depth analysis of how this policy is based on a micro-management approach that is embedded in a number of risk society preconceptions. The contradictions and potential lessons that we can conclude from the theory's application to the US precision strike policy will thus be discussed in this section. The chapter ends with a brief analysis of the main strategic and political implications that result from the Obama Administration's intensive use of drones in counter-terrorism operations, as well the potential consequences it might have for the future of precision strike operations.

Security and war in world risk society

Risk is an ontological condition of society (Beck, 2009). According to Ulrich Beck (2009: 40), security in late modernity is characterised by 'the collapse of ontological security' and with the ascendancy of the 'regime of non-knowing'. This happens when 'at least one factor in the classical security equation – agent, intention, potential – becomes an unknown' (ibid.: 40). In that regard, risk and security are associated in three different ways, in particular regarding the definition of dangers, time, and policies.

For any national security apparatus dangers are generally defined as threats. They 'can be precisely identified and measured on the basis of the capabilities an enemy has to realise a hostile intent' (Rasmussen, 2006: 1). This is in stark contrast with the volatility, unpredictability, and immateriality of risk. In dangers that fall within a security framework, such as terrorism or global pandemics, risk, or what is perceived as such, 'irremediably leaves behind the confines of the state system' (Kessler and Werner, 2008: 17). It is envisioned as what it might be, rather than as what it is. As stated by Christopher Coker (2009: 66):

> [r]isk is fundamentally a belief or attitude, a way of thinking about the world. To the extent that our social experience generates the suspicion that we are at risk most of the time we tend to feel even more anxious.

Rather than deriving our anxieties from the threats that we face, we construct dangers that are the ideational materialisation of our ontological condition of anxiety. In that sense, security policies informed by this risk condition do not aim to deal with a particular threat, but rather to reduce our anxiety level by managing a security concern. In the absence of the visibility of danger, this preventive move is defined and sustained by expert knowledge (Griner, 2002), reinforcing a trend identified by Rita Abrahamsen and Michael C. Williams (2010: 135) in which 'security becomes increasingly a question of the right technical solutions, and not a question of justice or social and political reform'. It is thus within this technical mind-set that risks are approached, not in order to be 'solved', but merely to be managed.

Time is another key difference between risk and threat: whereas the latter is framed in the present, the former is framed in the future. It is framed as a scenario that becomes politically meaningful when attached to specific policies that have the ultimate end of preventing risks from materialising themselves into real catastrophes (Rasmussen, 2006: 4). This leads to a precautionary approach (Kessler and Werner, 2008: 290) that, contrary to prudence, 'demands that we act under scientific and causal uncertainty' (Aradau and van Munster, 2007: 101), while attempting 'to provide the image of control where there is none' (Rasmussen, 2006: 37).

Security decisions are taken in a strategic framework informed by what Rasmussen (2001) has labelled as reflexive politics, i.e. the politics of reflexive (or late) modernity that is defined by the constant construction of meaning that rationalises risks and allows for them to be managed even though they are neither present nor necessarily real. In this context, 'pre-active politics is attached a high value, whereas re-active politics is devalued' (Beck, 2009: 41).

As a security tool, war becomes a technological endeavour, defined and undertaken by experts with ultimate risk management goals. Focusing on the Atlantic Alliance, Coker (2009: 24) argues that 'risk has become the organizing principle for allied decision making. Missions are set, tasks determined and national caveats put in place according to the risks involved'.

For Ulrich Beck, the wars of world risk society, i.e. risk wars, are a 'confused mixture of new and old wars, of virtual wars and nationally and transnationally operating terrorism' (Beck, 2009: 149). These wars involve, on the one hand, the practice of risk management and, on the other hand, that of risk transfer (Beck, 2009). Bluntly stated, it affects both the *jus ad bellum* and the *jus in bello* practices, particularly of Western countries. The former

> designates [...] military interventions in foreign (not hostile), more or less unstable (both collapsing and stable) states with the goal of minimizing and controlling a 'global risk' (transnational terrorism, the proliferation of atomic, chemical, and biological weapons of mass destruction, etc.).
>
> (Beck, 2009: 149)

The latter refers to Martin Shaw's (2005) argument that the Western world has been developing a strategic culture based on the redistribution of the risks of war, placing those risks almost exclusively on the populations and territories where they intervene, rather than on their own soldiers and people. The ultimate goal is to control risks and minimise (their own soldiers') casualties (Beck, 2009: 151).

In Beck's opinion, these wars are only acceptable to the Western public opinion as long as they fit the 'normality of Western societies and macroeconomies' (Beck, 2009: 152). Western democracies' tolerance for casualties is increasingly low (Sauer and Schörnig, 2012), which obliges political leaders to find risk-free technologies for waging war. Ultimately, the 'globality of risk does not mean the equality of risk' (Griner, 2002: 154). As argued by Coker (2009: 35), '[w]ar has been scaled down, as too has our strategic ambition; the West is no longer in the business of building a New World Order but managing the Global Disorder instead'. Accordingly, precision strikes are no longer used to win, but to manage; they are not part of a war effort, but the whole effort; they are not aimed at targets, but at individuals.

The increased use of precision strike technology, particularly through unmanned systems, is a consequence of this risk-averse mental framework. The tendency is for the progressive autonomisation and weaponisation of these systems (Sauer and Schörnig, 2012), with the consequent automatisation of warfare (see Coker, 2004; Lucas Jr., 2010; Singer, 2009). In this regard, the novelty and importance of drones is not inherent to their technological capacity, but in how they shape precision strikes. In particular, how their level of accuracy has led political leadership in the US to develop a micro-management attitude regarding their use: each strike, each mission, each killed insurgent results from a direct political order.

The control over the use of violence is no longer mediated by geographical distance or technological limitations.

After briefly addressing the evolution of US policy concerning precision strikes and the Obama Administration's use of UAVs, we will return to this issue in the third section of this chapter.

Precision strikes and US interventionist policy

The US has employed precision-guided weapons since the end of World War II[1] (Huiss, 2012). Throughout the Cold War advanced weapon systems such as precision-guided munitions, UAVs, and sensors were developed and many such systems were widely employed in the Vietnam War.[2] However, the exploitation of conventional guided weapons was limited during the Cold War due to their general inaccuracy, unreliability, and susceptibility to simple countermeasures (Watts, 2007). Operation Desert Storm in 1991 marked a watershed in the use of precision-guided munitions in American military strategy (Hayward, 2010; Watts, 2007). The integration of precision-guided munitions with other advanced technologies helped the US military to quickly and securely take out key strategic Iraqi targets (Mahnken, 2008).

The success of the bombing campaign in Iraq in the early 1990s revealed how precision weapons transformed the traditional logic of war (Hallion, 1997). The Gulf War provided a showcase for several new military technologies, such as precision-guided munitions and stealth, and led many commentators to announce the onset of a Revolution in Military Affairs. Throughout the remainder of the decade the US embraced these new military technologies as a panacea for dealing with its adversaries abroad. As Mahnken (2011: 50) has keenly observed

> ... the combination of stealth and precision-guided munitions gave US air forces the ability to strike adversaries from the air with near impunity. In addition, airpower seemed uniquely suited to the types of conflicts in which the United States was involved: wars for limited aims, fought with partial means, for marginal interests. Airpower coupled with PGMS (precision-guided munitions) appeared to offer the ability to coerce Iraq, intervene in the Balkans, and retaliate against terrorist groups while avoiding the difficult decisions associated with a sustained commitment of ground forces.

Precision weapons received yet another boost with Operation Allied Force, in 1999. The intervention in Kosovo revealed a new generation of precision weapons guided by the Global Positioning Systems (GPS), particularly the GBU-31 Joint Direct Attack Munition (JDAM).[3] The new military technologies used against Serbian air defence systems[4] and garrisons contributed to the wholesale increase in the number of precision-guided munitions used in combat operations (Huiss, 2012; Tilford Jr., 1999). The military intervention in Kosovo also bore witness to the use of modern UAVs in reconnaissance and surveillance missions[5] (Huiss, 2012).

The application of such state-of-the-art technology in the Balkans convinced many US officials that it was possible to wage war more efficiently and resourcefully (Cordesman, 2005; Ignatieff, 2001; Kagan, 2005). Enthusiasts argued that, in combination with new tactical concepts, the advantages were manifold. The capacity to strike targets accurately from a distance reduced many of the risks of conventional warfare. Large masses of soldiers could be removed from the battlefield, reducing the danger to them. Distance also reduced the possibility of enemy forces attacking US assets, as well as unshackling the constraints of securing bases and infrastructures abroad.

The proliferation of precision-guided weapons and associated technologies on the battlefield was in sync with the emerging military concepts associated with the Revolution in Military Affairs. Official American documents and studies stressed the need for the US to promote a doctrinal and technological overhaul in order to successfully prevail over the security and defence challenges of the future. For instance, the Joint Chiefs of Staff publication *Joint Vision 2010* argued that with 'precision targeting and longer range systems, commanders can achieve the necessary destruction or suppression of enemy forces with fewer systems, thereby reducing the need for time-consuming and risky massing of people and equipment' (Shalikashvili, 1996: 18).

The years that followed attested to a dramatic increase in precision-guided weapons, making them a new standard of warfare (Mahnken, 2011). Particularly noteworthy was the importance precision-guided weapons gained as a means of fighting terrorism.[6] For example, in 1998, the Clinton Administration used precision-guided weapons to answer the terrorist attacks on US embassies in Kenya and Tanzania. The US struck back by launching Tomahawk Land Attack Missiles against targets in Afghanistan and Sudan in Operation Infinite Reach. It is because President Clinton (1998) was speaking in and to a risk society that it was perfectly logical for him to justify the action as a 'strike at the network of radical groups affiliated with and funded by Osama bin Laden'. Targeting something as vague as a network of groups with something as concrete as a land missile could only be possible in a context in which it is necessary to deal materially with a less than precise danger.

As international terrorism gradually consumed US government officials' attention, the Clinton Administration increasingly regarded precision strikes as the best way of dealing with the impending threats. Accordingly, Clinton signed intelligence orders allowing for the use of lethal force against bin Laden and several of his senior lieutenants (Miller, 2001). Reluctant to place a large number of troops on the ground, the Clinton Administration favoured using precision strikes (particularly cruise missiles) to contend with Al-Qaeda. It also employed Predator unmanned aerial vehicles to collect detailed real-time video and photographs on the movements of top Al-Qaeda operatives (Miller, 2001).

Clinton's use of precision strikes to deal with many of the post-Cold War security challenges created a precedent for the years to come. By applying a tactic which sought to minimise resources and risks while simultaneously maximising military efficiency, the Clinton Administration paved the way for more robust interventions in the future. What began as a tactic for degrading an enemy state's military infrastructure and will to fight would develop into a full-fledged policy option for dealing with international terrorism.

The terrorist attacks of 11 September 2001 led the Bush Administration to use these new methods more audaciously (Byman, 2006). More precisely, the US's 'War on Terror' led to an adoption of targeted killings[7] as a means of fighting terrorism, particularly Al-Qaeda. Prior to this date, only Israel used this method explicitly in its fight against terrorism (Cullen, 2007; Statman, 2008). However, immediately after 11 September, President Bush authorised the CIA to use lethal drone strikes to disrupt and eliminate the Al-Qaeda network (Woodward, 2002; Yoo, 2011). As a result, the US began to target and kill terrorist operatives in countries such as Afghanistan, Iraq, Pakistan, and Yemen (see Blum and Heymann, 2010; Cullen, 2007; Murphy and Radsan, 2009; Statman, 2008; Williams, 2010).

Throughout the years, the programme increased its geographic scope and target list (Meyer, 2006) and targeted drone killings developed into one of the favoured means of combating terrorists when US forces were not openly employed on the ground[8] (Hudson *et al.*, 2011). This development was catalysed by the advances in satellite imagery, sophisticated electronic surveillance, UAVs, and precision-guided munitions, all of which conferred greater capacity and efficiency to the US in striking targets virtually anywhere and anytime. Despite the trenchant debate on the US's employment of targeted killings, particularly regarding its legality,[9] the Bush Administration generalised this tactic and institutionalised it through several policy documents (see Chairman of the Joint Chiefs of Staff, 2006; White House, 2006).

Barack Obama came to the White House promising to revise many of his predecessor's policies (Obama, 2008). The generalised discontent with the Bush Administration's handling of the wars in Afghanistan and Iraq was a political asset that the Obama campaign exploited effectively (Cassino and Erisen, 2010). In effect, George W. Bush left office as one of the most unpopular presidents in American history. According to a CNN/Opinion Research Corporation survey, at the end of his second term in office, 71 per cent of Americans disapproved of President Bush's management of the presidency (CNN, 2008b). Greatly contributing to this negative outlook was the US involvement in the War in Iraq. Five years after invading Iraq and ousting Saddam Hussein from power, most Americans were disgruntled with Bush's performance in handling the ensuing conflict[10] (CNN, 2008a).

Consequently, Obama (2007a, 2008) outlined a vast array of challenges for the US in order to rebuild its international leadership: ending the war in Iraq responsibly, focusing on global terrorism (particularly in Afghanistan and Pakistan), revitalising the military, halting the proliferation of nuclear weapons, rebuilding alliances/partnerships, and promoting democratic state-building. According to Obama (2007a: 7), while the US had to maintain the capability to defeat threats quickly and decisively, it 'must become better prepared to put boots on the ground in order to take on foes that fight asymmetrical and highly adaptive campaigns on a global scale'. More precisely, Obama (2008: 117) pledged to augment the number of US ground forces, namely by supporting 'plans to increase the size of the Army by 65,000 troops and the Marines by 27,000 troops'.

However, as mentioned above, President Obama continued his predecessor's policy of targeted killings.[11] In fact, the Obama Administration greatly increased the use of drone attacks. In his first week of office, Obama authorised two CIA Predator strikes in Pakistan against Al-Qaeda operatives (Murphy and Radsan, 2009). In the first year alone, the Obama Administration conducted more drone attacks than Bush did during his entire presidency (Shane, 2009). The use of precision drone strikes against Al-Qaeda leaders was judged 'very effective' by US officials such as the former CIA Director Leon Panetta (CNN, 2009). Members of other groups, such as Taliban and Haqqani Network militants, have also been targeted and eliminated using drone strikes (Mehsud, 2013; Shane, 2009). Moreover, Obama's former Deputy National Security Advisor for Homeland Security and Counterterrorism and current CIA director John Brennan stated that besides being legal and ethical, drone strikes were a wise option due to the fact that they provided a low risk for US personnel and civilians (Brennan, 2012).

At the operational level, the use of drones for targeted killings guarantees a sustained persistence over potential targets and near-instantaneous responsiveness (Zenko, 2013). The former results from the drones' capacity to remain in the air for over three times the number of hours conventional combat aircraft can.[12] The latter allows for radically shrinking the 'find-fix-finish' loop, contributing to greater efficiency in eliminating targets. In other words, while traditional precision strike weapons, such as cruise missiles, required several hours of prior notice in order to evaluate the intelligence information, obtain authorisation, programme the weapon platform, and strike the target, the use of drones allows for near-instant strikes. This immediacy allegedly lessened the collateral damage resulting from the precision strikes.

Targeted killings are not only conducted by unmanned armed drones. Obama's greatest accomplishment in fighting terrorism so far involved the killing of bin Laden by the JSOC's Naval Special Warfare Development

Group (a.k.a. DevGru) on a 'kill/capture' mission in May 2011 (Ambinder, 2011). Nonetheless, the overwhelming majority of all non-battlefield targeted killings since 11 September 2001 have been conducted by drones[13] (Zenko, 2013). Moreover, the frequency and intensity of drone attacks has significantly increased in the most recent years. Since 2009 the Obama Administration has launched over six times as many drone strikes as the Bush Administration (Boyle, 2013), leading to over ten times as many deaths. While data varies according to the different sources, available estimates all point to a manifest increase in the number of drone strikes, as well as killings.[14] For instance, in the Yemeni case, despite rebutting any official involvement, through its drone attacks on Al-Qaeda and other Islamic groups the US has become directly involved in the internal conflict (Boyle, 2013). In Yemen, the US has conducted about 50 strikes, killing anywhere from nearly 400 to over 799 individuals (Zenko, 2013).

This course of action is in line with President Obama's own statements halfway through his first term. In his address to the nation explaining the US-led intervention in Libya, Obama (2011) put forward the principal tenets of the Obama Doctrine: assessment of national interest, American leadership, limited role of US forces, broad coalition, international mandate, and division of labour between allies. Particularly important was the focus on limited US military involvement. In his speech, the President stated several times that the US would not put troops on the ground. Rather, the US would 'focus our unique capabilities on the front end of the operation', transferring other duties to allies and partners (Obama, 2011). The US's unique contribution was eliminating Qaddafi's military assets through its sophisticated and devastating precision strike capabilities.

Accordingly, rather than placing troops on the ground the Obama Administration has preferred to use its technological edge to fight Al-Qaeda and other international terrorist groups. For most US officials, drone strikes are a highly efficient means of increasing the government's chances of crushing insurgencies, degrading terrorist organisations' capabilities, curtailing insurgent attacks, and reducing overall levels of violence (Boyle, 2013; Johnston, 2012), whilst concurrently decreasing the risks to US forces and assets, as well as civilians. Additionally, they also serve to satisfy domestic demands for a strong response to terrorism (Hajjar, 2012) without having to shoulder additional commitments of employing forces on the ground. Yet, this rationale results from the assumption that targeted killings are directed at high-level Al-Qaeda leaders or other high-value targets who pose a serious threat ('not speculative') to the US and who cannot be captured (Obama, 2012). However, the bulk of deaths resulting from drone strikes have been 'low-level, anonymous suspected militants who were predominantly engaged in insurgent or terrorist operations against their governments, rather than in active international terrorist plots' (Zenko, 2013: 10).

Another innovative aspect of the Obama Administration's targeting policy is the growth of 'signature' strikes.[15] Whereas the Bush Administration focused predominantly on 'personality' strikes – i.e. named, purportedly high-value leaders of armed non-state organisations – under Obama targets are defined increasingly on 'profiles' based on behavioural patterns (Stanford/NYU, 2012). This authorises strikes against unidentified individuals or groups of individuals who reveal certain defining characteristics related with terrorist activity. According to existing sources, in recent times the majority of CIA drone strikes have been signature strikes (Heller, 2013). In addition, the Administration has also begun targeting US citizens. The deliberate listing of citizens in the JSOC kill list reportedly began with Anwar al-Awlaqi[16] and has been followed by others (Hajjar, 2012).

The definition of the targets is conducted at the highest level of the decision making apparatus (Becker and Shane, 2012). The process involves a weekly secure video teleconference managed by the Pentagon in which over 100 government officials analyse the biographies of suspected terrorists and subsequently submit a list of targets to the President. The approval of the targets can be extended throughout several sessions until the 'nominations' are finally submitted to the White House 'where by his own insistence and guided by Mr. Brennan (his counterterrorism adviser), Mr. Obama must approve any name. He signs off on every strike in Yemen and Somalia and also on the more complex and risky strikes in Pakistan – about a third of the total' (Becker and Shane, 2012).

As Daniel Klaidman's (2012) account of the decision making process attests, the target assessment results from a fraught debate between senior officials. Besides top CIA and military officials, the President and his senior advisors also rely on the legal counsel of the top lawyers of the different agencies. As potential targets are presented, the different participants scuffle to try to convince the President of the usefulness or ineffectiveness of eliminating them. Throughout the process, Obama questions his advisors on the efficacy of each strike. Above all, the President tries to evaluate how each particular strike can assure 'the safety of America on the one hand and the righteousness of what America stands for on the other' (Klaidman, 2012). The need to limit additional involvement in internal conflicts abroad guides the President's judgment all the way through the decision process.

The Obama Administration's wholesale use of targeted-drone killings has spurred significant criticism at home and abroad. Dispute has grown over multiple issues such as the legality, oversight, effectiveness, and risks resulting from drone killings. The following section will examine these arguments, paying particular attention to how these target killings can be framed in the context of risk society.

Risk transference and the micro-management of war

As argued above, according to Ulrich Beck (2009), risk transference is one of the two elements present in risk wars. Western risk societies are not particularly tolerant of war casualties. They are 'safety' societies, 'permanently on the defensive' (Coker, 2009: 26), thus favouring approaches to war that spare the life of their own soldiers. In that sense, it is not so much the war and violence that is disturbing to these societies; it is the proximity and the visibility of it. As pointed out by Beck (2009: 153),

> '(l)egitimate' and 'successful' is ultimately the zero-deaths war – correction, the zero-*own*-deaths war – which blots out the countless deaths on the other side, in particular among the civilian population, thereby rendering the war virtual for the Western nations and all the more brutally real for the victim populations.

This 'invisibilisation' of war takes place as labour (soldiers) is replaced by capital (technology) (Sauer and Schörnig, 2012: 369), making war all the more lethal, albeit less human (see Coker, 2004). In the US case, drones are the weapons of choice for this transfer process. It is through them that Washington has sustained, and in some cases expanded, its counter-terrorism policy without generating a significant political setback at home.

UAVs and UCAVs (unmanned combat air vehicles) are integrated into complex technological systems which involve countless numbers of personnel to control, manage, and maintain each of these unmanned aircrafts. In that sense, they can also be considered labour-intensive. However, this labour is risk-free because it does not directly engage in fighting and is physically immune to the outcome of the missions on the ground. Throughout history, technology has contributed to make warfare more precise, more efficient, and more lethal. Its implementation, though, has always required the sacrifice of human lives. With drones we have almost reached the point of full replacement, in which no single person has to be sacrificed in order to effectively complete a military mission. The human investment that has always been necessary for waging wars is thus now replaced by technological development: capital replaces labour.

As mentioned above, the immense risks associated with the military missions in Afghanistan and Iraq and that contributed to the erosion of George W. Bush's popularity have been progressively replaced by the Obama doctrine that focuses on surgical Special Forces' missions, combined with precision strikes wherever significant terrorist activity has been detected.

The key issue here is exactly how those activities are signalled and approached by the US Administration. If drones are the privileged instruments, targeted killing is the privileged practice. Putting to one side the ethical and legal issues associated with this type of practice, when put against the world risk society framework, there are three main problems associated with this security approach.

A first problem is related to the necessary materialisation of risks, and their consequent individualisation in a war risk context. Precision strikes can only be used against defined targets. Overcoming the 'anonymity of risks' (Beck, 2009: 41) thus implies the precautionary attribution of agency, i.e. the construction of a defined threat from an undefined risk. In the case of terrorism this is done by identifying, and consequently targeting, a number of individuals who are considered to be part of this nebulous concept of global terrorism, in most instances (but not exclusively) associated with Al-Qaeda. Often, this association is only loosely established and the information from which it is based is not entirely reliable (Kessler and Werner, 2008: 291). From a risk society perspective the inconsistency of this process is noticeable. Precision strikes are used as a form of risk-avoidance, but their use necessarily demands the materialisation and individualisation of undefined risks (as we saw above in President Clinton's justification of missile strikes against targets in Afghanistan and Sudan). The specificity of the weapon conflicts with the vagueness of the target, which implies that in order for both to be compatible there is a necessary degree of arbitrariness in the whole process.

In order to reduce this arbitrariness the Obama Administration has adopted a micro-management approach to target killings – which we take as the second problem associated with the US targeted killing policy. This happens for two reasons, which are both important in the general assessment of the US's precision strikes policy: first, due to the limited number of people involved in the decision making process and, second, to the 'proximity' between the action on the ground and the decision making process. The ability of decision makers to see the implementation of their decisions in real time is certainly a novelty with few comparable examples in the history of warfare. Some of those decisions are merely based on the movement or characteristics of a given group – the so-called signature strikes – recognised as highly inaccurate by some officials from the State Department (Becker and Shane, 2012).

The second problem results from the 'management' disposition of the whole endeavour. Unlike the rudimentary strikes of World War II, the contemporary use of combat drones does not aim at winning a war, for the war against terrorism that the US is waging does not have a defined territory, a clear enemy, or a clear end. It is an endless risk (micro-) management exercise.

Finally, a third problem has to do with the effects of the policy of precision strikes. Again, if we follow a risk perspective, then the idea of '(s)ecurity and control as understood in first modernity are impossible' (Griner, 2002: 153), since '(e)very diagnosis, every proposal is subject to the laws of the multiplication of risk which the critic is trying to counter'

(Beck, 2009: 143). This impossibility means that any policy that attempts to deal with existing risks will always generate new risks. The use of drones for targeting individuals in countries against which the US is not at war is no exception to this. Reactions to this might be felt at three different levels. First, the countries where the strikes take place might feel aggrieved for the violation of sovereignty by the US. The operation that led to the capture and killing of Osama bin Laden in Pakistan was angrily received by both the Pakistani political and military elites (Crilly, 2011). Second, the groups that are targeted might, as a consequence of the limited timing available for acting, resort to less precise attacks, resulting in a higher number of innocents killed. Finally, the communities where precision attacks take place might be increasingly radicalised against the US. Not only might this create privileged recruitment areas for extremist movements (Sauer and Schörnig, 2012: 373), it might also negatively affect any potential US interests in the regions where this type of strike is conducted. The potential blowbacks (Davies, 2012) of these operations are not possibly calculated to their full extent, since they are unintended risks of policies that attempt to address other risks (terrorism) as if they were precise threats with precision strike weapons – this while, simultaneously, transferring the burden of the exercise (risk) to the peripheral zones of the international system. In the end, '[t]he strong suffer with the weak because in reflexive society no one can escape the consequences of their actions' (Rasmussen, 2006: 40).

Conclusion

This chapter has attempted to address the political and strategic consequences of the US's current precision strike policy from a 'risk society' perspective. As analysed, this policy is based on the intensive use of unmanned systems against individuals who are classified as relevant terrorist operatives according to lists previously approved by the top echelon of the US Administration and implemented in a largely secretive manner. Based on Beck's, Coker's, and Rasmussen's works on risk wars, we can conclude that the US precision strike policy is fraught with major inconsistencies derived from both the rationale and method of precision strikes, namely: the attempt to materialise and individualise risks; the attempt to provide a sense of control – through micro-management – of what cannot be controlled; the belief in risk-free wars (Rasmussen, 2006: 43); and, finally, the consequent unintended risks that derive from the transference of the war burden to the communities where the strikes take place.

It is highly unlikely the Obama Administration will change its stance on precision strikes and its intensive use in its counter-terrorist policy during its second term. It is however expected that those missions will acquire greater public visibility and that the definition and execution of the so-called target lists are, due to public pressure, made according to a more precise methodology. In particular legal and ethical considerations, as well as the strategic consequences of precision strikes, will increasingly determine the Administration's engagement in risk management warfare.

As the number of UAV strikes increase, management of the process may increasingly be relegated to non-political actors. Moreover, the continued developments in robotic technology and artificial intelligence pose the risk of the emergence of automated strikes in the near future (Keller, 2013). This should raise serious alarms for US decision makers who should bear in mind Rupert Smith's (2008: 215) sensible caveat that 'it is crucial that military objectives should be chosen for their value in achieving the political purpose or aim, not just because they are militarily possible'.

However, US decision makers will continually seek to maintain potential threats and risks at a distance and deal with them as rationally as possible. In American foreign policy moral and legal imperatives have usually given way to immediate political considerations. Historical record testifies to this fact. The future will most certainly confirm this tendency.

Notes

1 Some of the first guided American weapons were the V-1 cruise missile, the V-2 ballistic missile, and the Mark-24 acoustic-homing torpedo (Huiss, 2012; Watts, 2007).
2 For instance, the US Air Force and Navy expended over 28,000 laser-guided bombs in South East Asia between 1968 and 1973 – mostly against bridges and transportation checkpoints (Mahnken, 2008).
3 The JDAM is composed of a guidance kit that allows for the transformation of unguided bombs into all-weather 'smart' munitions, which are guided by an integrated inertial guidance system coupled to a GPS receiver.
4 The air defence systems were essentially command and control facilities, airfields, radars, and surface-to-air missile sites.
5 Namely the Air Force RQ-1A Predator.
6 Wedgwood (1999) identifies four main strategies that the US adopted for responding to terrorism prior to 11 September 2001: (1) use the tools of criminal justice; (2) treaty agreements to establish new international norms and enforcement mechanisms; (3) disrupt terrorist structures through civil sanctions; and (4) prudent use of military force to prevent terrorist attacks and to degrade terrorist infrastructures.
7 There are a wide variety of terms used to describe this tactic, e.g. preventive killings, active self-defense, extrajudicial killings, interceptions, and named killings. However, the term 'targeted killing' has become the most broadly applied and accepted, certainly due to the fact that it does not presume any form of approval or disapproval of the tactic (Cullen, 2007). For the current paper we use targeted killing to refer to intentional slaying of a specific individual or group of individuals undertaken with explicit approval of governmental decision makers (David, 2003).
8 The weapon of choice for carrying out targeted killings is the MQ-1 Predator drone (UAV) developed by General Atomics Aeronautical in conjunction with DARPA, armed with Hellfire missiles.
9 For insight into the different arguments exchanged concerning the legality of targeted killings see, for example, Blum and Heymann (2010), Cullen (2007), Gross (2006), Murphy and Radsan (2009), Shah (2010).

10 Only 31 per cent of Americans approved of how President Bush was handling the War according to a CNN poll released in March 2008.
11 In reality, Obama never openly condemned Bush's tactic of drone killings. As a matter of fact, during a 2007 campaign speech at the Woodrow Wilson International Center, Obama sanctioned targeted killings when referring to Al-Qaeda activities in Pakistan (Obama, 2007b).
12 The armed drones most commonly used – i.e. Predator and Reaper – can remain in the air for over 14 hours, while aircraft such as the F-16 or A-10 can only wait for four hours or less (Zenko, 2013).
13 More precisely, 95 per cent of non-battlefield targeted killings result from drone strikes, while the remaining 5 per cent result from JSOC raids and AC-130 gunships, as well as offshore sea- or air-launched cruise missiles (Zenko, 2013).
14 For estimates on drone strikes and killings see the Bureau of Investigative Journalism (www.thebureauinvestigates.com/category/projects/drones) and the New American Foundation (http://counterterrorism.newamerica.net/drones).
15 The signature strikes are also designated terrorist-attack-disruption strikes (TADS).
16 Al-Awlaqi was an American-born Muslim cleric residing in Yemen who was characterised as a leader of Al-Qaeda in the Arabian Peninsula and fell to a drone attack on 30 September 2011.

References

Abrahamsen, R. and Michael, C. W. (2010) *Security beyond the State: Private Security in International Politics*, Cambridge: Cambridge University Press.

Ambinder, M. (3 May 2011) 'The Secret Team that Killed bin Laden', *National Journal*, www.nationaljournal.com/whitehouse/the-secret-team-that-killed-bin-laden-20110502 (accessed 12 January 2013).

Aradau, C. and van Munster, R. (2007) 'Governing Terrorism through Risk: Taking Precautions, (Un)Knowing the Future', *European Journal of International Relations*, 13, 1: 89–115.

Beck, U. (1992) *Risk Society: Towards a New Modernity*, London: Sage.

Beck, U. (1998) 'Politics of Risk Society', in Franklin, J. (ed.) *The Politics of Risk Society*, Cambridge: Polity Press, pp. 9–22.

Beck, U. (2009) *World at Risk*, Cambridge: Polity Press.

Becker, J. and Shane, S. (29 May 2012) 'Secret "Kill List" Proves a Test of Obama's Principles and Will', *New York Times*, www.nytimes.com/2012/05/29/world/obamas-leadership-in-war-on-al-qaeda.html?pagewanted=all&_r=0 (accessed 23 January 2013).

Bergen, P. and Tiedmann, K. (2011) 'Washington's Phantom War: The Effects of the US Drone Program in Pakistan', *Foreign Affairs*, 90, 4: 12–18.

Blum, G. and Heymann, P. (2010) 'Law and Policy of Targeted Killing', *Harvard National Security Journal*, 145, 1: 145–70.

Boyle, M. (2013) 'The Costs and Consequences of Drone Warfare', *International Affairs*, 89, 1: 1–29.

Brennan, J. (2012) *The Ethics and Efficacy of the President's Counterterrorism Strategy*, Remarks at the Wilson Center Conference on 'The Efficacy and Ethics of U.S. Counterterrorism Strategy' (30 April 2012), www.wilsoncenter.org/event/the-efficacy-and-ethics-us-counterterrorism-strategy (accessed 13 January 2013).

Byman, D. (2006) 'Do Targeted Killings Work?', *Foreign Affairs*, 85, 2: 95–111.

Cassino, D. and Erisen, C. (2010) 'Priming Bush and Iraq in 2008: A Survey Experiment', *American Politics Research*, 38, 2: 372–94.

Chairman of the Joint Chiefs of Staff (2006) *National Military Strategic Plan for the War on Terrorism*, Washington, D.C.: Department of Defense.

Clinton, W. (1998) 'Address to the Nation on Military Action against Terrorist Sites in Afghanistan and Sudan', 20 August 1998, www.presidency.ucsb.edu/ws/index.php?pid=54799 (accessed 4 January 2013).

CNN Poll (2008a) 'Bush's Popularity Hits New Low', http://articles.cnn.com/2008–03–19/politics/bush.poll_1_iraq-war-approval-number-approval-rating?_s=PM:POLITICS (accessed 7 October, 2010).

CNN Poll (2008b) 'More Disapprove of Bush than any other President', http://articles.cnn.com/2008–05–01/politics/bush.poll_1_disapproval-rating-new-poll-polling-director?_s=PM:POLITICS (accessed 7 October, 2010).

CNN (2009) 'U.S. Airstrikes in Pakistan Called "Very Effective"' (18 May 2009), http://edition.cnn.com/2009/POLITICS/05/18/cia.pakistan.airstrikes (accessed 11 January 2013).

Coker, C. (2004) *The Future of War: The Re-Enchantment of War in the Twenty-First Century*, Victoria, Australia: Blackwell Publishing.

Coker, C. (2009) *War in an Age of Risk*, Cambridge: Polity Press.

Cordesman, A. (2005) 'The Strategic Impact and Military Effectiveness of the Air and Missile Campaign', in Bolt, P., Coletta, D., and Shackelford Jr., C. (eds) *American Defense Policy*, Baltimore: The Johns Hopkins University Press, pp. 360–4.

Crilly, R. (2011) 'Osama Bin Laden Dead: Angry Pakistan Drops Intelligence Sharing With West', *The Telegraph* (14 May 2011), www.telegraph.co.uk/news/worldnews/al-qaeda/8513920/Osama-bin-Laden-dead-angry-Pakistan-drops-intelligence-sharing-with-West.html (accessed 11 January 2013).

Cullen, P. (2007) *Role of Targeted Killing in the Campaign against Terror*, Carlisle, PA: US Army War College.

David, S. (2003) 'Israel's Policy of Targeted Killing', *Ethics and International Affairs*, 17, 1: 111–26.

Davies, N. J. S. (2012) 'Obama's America: Waiting for Blowback', *Huffington Post* (28 August 2012), www.huffingtonpost.com/nicolas-j-s-davies/obama-foreign-policy_b_1831503.html (accessed 9 January, 2013).

Gates, Robert (2010) *The Report of the Quadrennial Defense Review*, Washington, D.C.: Department of Defense.

Giddens, A. (1998) 'Risk Society: The Context of British Politics', in Franklin, J. (ed.) *The Politics of Risk Society*, Cambridge: Polity Press, pp. 23–34.

Griner, S. (2002) 'Living in a World Risk Society: A Reply to Mikkel V. Rasmussen', *Millennium – Journal of International Studies*, 31, 1: 149–60.

Gross, M. (2006) 'Assassination and Targeted Killing: Law Enforcement, Execution or Self-Defence?', *Journal of Applied Philosophy*, 23, 3: 323–35.

Hajjar, L. (2012) 'Anatomy of the US Targeted Killing Policy', *Middle East Report*, 42, www.merip.org/mer/mer264/anatomy-us-targeted-killing-policy (accessed 4 January 2013).

Hallion, R. (1997) *Precision Guided Munitions and the New Era of Warfare*, Australia: Air Power Studies Centre.

Hayward, J. (2010) 'Air Power: The Quest to Remove Battle from War', in

Kassimeris, G. and Buckley, J. (eds) *The Ashgate Research Companion to Modern Warfare*, Surrey: Ashgate Publishing Company, pp. 49–72.

Heller, K. (2013) '"One Hell of a Killing Machine": Signature Strikes and International Law', *Journal of International Criminal Justice*, 11, 1: 89–119.

Hudson, L., Owens, C., and Flannes, M. (2011) 'Drone Warfare: Blowback from the New American Way of War', *Middle East Policy*, 18, 3: 122–32.

Huiss, R. (2012) *Proliferation of Precision Strike: Issues for Congress*, Washington, D.C.: Congressional Research Service.

Ignatieff, M. (2001) *Virtual War: Kosovo and Beyond*, New York: Picador.

Indyk, M., Lieberthal, K., and O'Hanlon, M. (2012) 'Scoring Obama's Foreign Policy: A Progressive Pragmatist Tries to Bend History', *Foreign Affairs*, 91, 3: 29–43.

Johnston, P. (2012) 'Does Decapitation Work? Assessing the Effectiveness of Leadership Targeting in Counterinsurgency Campaigns', *International Security*, 36, 4: 47–79.

Kagan, F. (2005) 'The Art of War', in Bolt, P., Coletta, D., and Shackelford Jr., C. (eds) *American Defense Policy*, Baltimore: Johns Hopkins University Press, 229–37.

Keller, B. (2013) 'Smart Drones', *New York Times* (16 March 2013), www.nytimes.com/2013/03/17/opinion/sunday/keller-smart-drones.html?_r=0 (accessed 15 October 2013).

Kessler, O. and Werner, W. (2008) 'Extrajudicial Killing as Risk Management', *Security Dialogue*, 39, 2–3: 289–308.

Klaidman, D. (2012) 'Drones: How Obama Learned to Kill', *The Daily Beast* (28 May 2012), www.thedailybeast.com/newsweek/2012/05/27/drones-the-silent-killers.html (accessed 23 January 2012).

Lindsay, J. (2011) 'George W. Bush, Barack Obama and the Future of US Global Leadership', *International Affairs*, 87, 4: 765–79.

Lucas Jr., G. (2010) 'Postmodern War', *Journal of Military Ethics*, 9, 4: 289–98.

Mahnken, T. (2008) *Technology and the American Way of War since 1945*, New York: Columbia University Press.

Mahnken, T. (2011) 'Weapons: The Growth and Spread of the Precision-Strike Regime', *Dædalus*, 140, 3: 45–57.

Mehsud, I. (2013) 'Suspected Drone Strike Kills 4 in Pakistan', *New York Times* (30 September 2013), www.nytimes.com/2013/10/01/world/asia/suspected-drone-strike-kills-4-in-pakistan.html?adxnnl=1&adxnnlx=1381919508-uMmTUSSBW-FASmh18KG6GMQ (accessed 15 October 2013).

Meyer, J. (2006) 'CIA Expands Use of Drones in Terror War', *Los Angeles Times* (29 January 2006), http://articles.latimes.com/2006/jan/29/world/fg-predator29 (accessed 8 January 2013).

Miller, J. (2001) 'Many Say the US Planned for Terror but Failed to Take Action', *New York Times* (30 December 2001), www.nytimes.com/2001/12/30/national/30TERR.html?pagewanted=all (accessed 4 January 2013).

Murphy, R. and Radsan, A. (2009) 'Due Process and Targeted Killing of Terrorists', *Cardozo Law Review*, 31, 2: 405–50.

Obama, B. (2007a) 'Renewing American Leadership', *Foreign Affairs*, 86, 4: 2–16.

Obama, B. (2007b) 'Obama's Speech at Woodrow Wilson Center (1 August 2007), www.cfr.org/us-election-2008/obamas-speech-woodrow-wilson-center/p13974 (accessed 13 January 2013).

Obama, B. (2008) *Change We Can Believe In: Barack Obama's Plan To Renew America's Promise*, New York: Three Rivers Press.
Obama, B. (2011) 'Remarks by the President in Address to the Nation on Libya' (28 March 2011), www.whitehouse.gov/the-press-office/2011/03/28/remarks-president-address-nation-libya (accessed 12 January 2013).
Obama, B. (2012) 'Obama Interview with Jessica Yellin on CNN', www.youtube.com/watch?v=edrjmULn2Jk (accessed 12 January 2013).
Rasmussen, M. (2001) 'Reflexive Security: NATO and International Risk Society', *Millennium – Journal of International Studies*, 30, 2: 285–309.
Rasmussen, M. (2006) *The Risk Society at War: Terror, Technology and Strategy in the Twenty-First Century*, Cambridge: Cambridge University Press.
Sauer, F. and Schörnig, N. (2012) 'Killer Drones: The "Silver Bullet" of Democratic Warfare?', *Security Dialogue*, 43, 4: 363–80.
Shah, S. (2010) 'War on Terrorism: Self Defense, Operation Enduring Freedom, and the Legality of US Drone Attacks in Pakistan', *Washington University Global Studies Law Review*, 9, 1: 77–129.
Shalikashvili, J. (1996) *Joint Vision 2010*, Washington, D.C.: US Government Printing Office.
Shane, S. (2009) 'CIA to Expand Use of Drones in Pakistan', *New York Times* (3 December 2009), www.nytimes.com/2009/12/04/world/asia/04drones.html?_r=0&pagewanted=all (accessed 12 January 2013).
Shaw, M. (2005) *The New Western Way of War*, Cambridge: Polity Press.
Singer, P. (2009) *Wired for War*, London: Penguin.
Smith, R. (2008) *The Utility of Force: The Art of War in the Modern World*, New York: Vintage Books.
Stanford/NYU (Stanford International Human Rights and Conflict Resolution Clinic and Global Justice Clinic at NYU School of Law) (2012) *Living Under Drones: Death, Injury, and Trauma to Civilians from US Drone Practices in Pakistan*, http://livingunderdrones.org/wp-content/uploads/2012/10/Stanford-NYU-LIVING-UNDER-DRONES.pdf (accessed 12 January 2013).
Statman, D. (2008) 'Targeted Killings', *Theoretical Inquiries in Law*, 5, 1: 179–98.
Tilford Jr., E. (1999) 'Operation Allied Force and the Role of Air Power', *Parameters*, 29, 4: 24–38.
Watts, B. (2007) *Six Decades of Guided Munitions and Battle Networks: Progress and Prospects*, Washington, D.C.: Center for Strategic and Budgetary Assessments.
Wedgwood, R. (1999) 'Responding to Terrorism: The Strikes against Bin Laden', *Yale Journal of International Law*, 24: 559–576.
White House (2003) *National Strategy for Combating Terrorism*. Washington, D.C.: The White House.
White House (2006) *National Strategy for Combating Terrorism*. Washington, D.C.: The White House.
Williams, B. (2010) 'The CIA's Covert Predator Drone War in Pakistan, 2004–2010: The History of an Assassination Campaign', *Studies in Conflict and Terrorism*, 33: 871–92.
Woodward, B. (2002) *Bush at War*, New York: Simon & Schuster.
Yoo, J. (2011) 'Assassination or Targeted Killings after 9/11?', *New York Law School Law Review*, 56: 57–79.
Zenko, M. (2013) *Reforming US Drone Strike Policies*, New York: Council on Foreign Relations.

2 European precision strike capabilities

A neoclassical realist perspective

Tom Dyson

Introduction

The post-Cold War defence reforms of the West European Great Powers (Britain, France, and Germany) have been characterised by a partial and selective emulation of the US-led Revolution in Military Affairs (RMA). However, significant differentiation exists in the pace of European emulation. Such differentiation in the defence reforms of Europe's three major military players has very important implications for the capacity of European states to 'burden-share' within NATO and for the ability of the EU to undertake autonomous stand-off precision strike operations through its Common Security and Defence Policy (CSDP). This chapter will analyse the reforms which have taken place to the force structures, military doctrines, and capabilities of the West European Great Powers and assess the relative capacity of these states to undertake the precision strike, networked operations which are associated with the RMA. It finds that while Britain and France have been regional pacesetters in emulation of the RMA, the German *Bundeswehr* has fallen behind on all three aspects of emulation: force structures, military capabilities, and military doctrine.

The chapter will seek to explain these patterns of convergence and divergence in RMA emulation through the use of Neoclassical Realist international relations theory. It argues that systemic variables – notably the balance of power – are driving convergence between all three states in the instruments and art of warfare. Domestic variables, particularly the autonomy of the core executive in defence policy, play an important intervening role in determining the scope and pace of convergence. The chapter finds that the high-level of 'executive autonomy' in the UK and France has permitted timely adherence to the dictates of systemic imperatives and military 'best-practice'. However, low executive autonomy in Germany has incentivised policy leaders to manage the temporality of convergence with the dictates of the contemporary security and operational environment.

The first section of the chapter analyses the US-led Revolution in Military Affairs and the content and timing of the partial and selective emulation of RMA undertaken by Britain, France, and Germany, focusing in particular on reforms to military structures, capabilities, and doctrines. The chapter then examines the utility of competing cultural and Realist frameworks for understanding the empirical patterns identified. It concludes by identifying the implications of the chapter's findings for policy makers in the three countries, as well as the key future challenges facing Britain, France, and Germany in developing precision strike capabilities.

Europe's selective emulation of the RMA

The RMA is characterized by three key dimensions: first, the creation of professional expeditionary forces with joint command structures (Reynolds, 2006: 458); second, Network-Centric Warfare that involves a move away from weapons platforms to knowledge-empowered networked forces which are capable of applying attritional force with agility and precision (Network-Centric Warfare (NCW) (Dahl, 2002: 5) through both manned and unmanned weapons systems; the third feature of the RMA is doctrinal and has cohered around the concept of Effects-Based Operations (EBO) (Farrell, 2008: 779). EBO are characterized by the use of networked activities and the mobilisation of all sources of national power (political, economic, military, and diplomatic) to attain strategic effects against near-peer competitors (Farrell, 2008: 779).

Although the RMA remains at the forefront of US defence reform, since 2006 a much stronger emphasis has been placed upon preparing for stabilization and counterinsurgency operations. The RMA has therefore adapted to focus not only upon the utility of new technologies in operations against near-peer competitors, but on how technology (particularly UCAVs) can be harnessed in operational contexts such as Stabilisation and Counterinsurgency, where conflict varies quickly in intensity across the conflict spectrum (Dyson, 2010: 23–7). In its emphasis upon the importance of a population-focused approach, the 2006 US Army/Marines Counterinsurgency (COIN) Doctrine Field Manual formed an important shift away from EBO's focus on kinetic operations (Hauser, 2007: 167). EBO were also removed from US Joint Doctrine in 2008, while the 2009 US defence budget focused investment around RMA and other capabilities more suited to irregular conflict (Dyson, 2010: 25–6).

Britain and France: timely adaptation to the RMA

British and French defence reforms have involved a partial and selective emulation of the RMA. The first feature of this emulation is the restructuring of command structures and development of joint, expeditionary forces; a process that began in France in 1995/96 and in Britain in 1997/98 (Dyson, 2008). Both the UK and France have sought to enhance the deployability, interoperability, and jointness of their armed forces. The

UK's 2010 Strategic Defence and Security Review (SDSR) plans to develop the capacity to deploy a one-off intervention force of 30,000 from a total force of 154,000 troops. Alternatively it proposes the capacity to undertake (concurrently) a long-term stabilisation operation of 6,500, a short-term complex intervention of up to 2,000 personnel as well as a short-term simple intervention of up to 1,000 troops (SDSR, 2010: 19). French defence planning assumptions aim (from 225,000 troops) to simultaneously deploy up to 15,000 troops in a major expeditionary operation and 7,000 troops undertaking crisis-management operations in up to three theatres (DWP, 2013: 4–5).

Second, during the early-mid 2000s, the British and French began to develop a Network-Enabled Capability (NEC). NEC seeks to exploit technology's tactical and operational advantages, but is more wary about networking's potential to transform warfare and deliver strategic effects (Farrell, 2008: 786–7). NEC has two dimensions. First, a technical dimension, involving the procurement of command, control, communications, computers, intelligence, surveillance, target acquisition, and reconnaissance (C4ISTAR) systems, including UAVs. The acquisition of C4ISTAR systems has been accompanied by the development of extensive manned and unmanned precision strike aerial systems. Both the UK and France have substantial UCAV capabilities and the UK has been particularly active in using its Reaper medium and high-altitude drones in Afghanistan to target enemy combatants (Baker *et al.*, 2013). The second dimension of NEC is doctrinal: the reconfiguration of thinking on command and control in the context of the introduction of these new technologies. British and French doctrinal development in the context of the RMA has cohered around the Effects Based Approach to Operations (EBAO).

In contrast to EBO, which focus on kinetic effects against near-peer competitors, EBAO recognises that military operations themselves have not been transformed, but instead that the approach to operations has changed. Embedded within the Comprehensive Approach (a multi-agency, cross-government approach to the planning and execution of operations), EBAO can enable the integration of all agencies of government in operations across the spectrum of intensity. Since EBO's decline, EBAO has given way to an emphasis on 'Effects-Based Thinking' that avoids the determinism of EBAO, while recognising the utility of EBAO in targeting infrastructure. In addition, UK and French defence reform has also focused upon the development of a significant body of doctrine and training to prepare its forces for Stabilisation and Counterinsurgency operations (Alderson, 2010; De Durand, 2010).

Germany: slow and reactive emulation of the RMA

German emulation of the RMA has mirrored that of the UK and France, but with significant delay. First, through reforms to command structures,

Germany has focused upon improving the jointness, deployability, and interoperability of the *Bundeswehr*. However, significant problems remain. Until 2010, the Bundeswehr continued to rely on conscription as a key source of military manpower. This is a form of recruitment that is manifestly unsuited to the challenges of the post-Cold War security environment, which necessitate highly trained soldiers, capable of dealing with conflict of rapidly varying intensity.

Furthermore, the *Bundeswehr* has a more limited capability than the British and French to deploy significant numbers of troops on overseas missions. Although the 2006 DWP outlined the intention of developing the capacity to deploy 14,000 troops abroad (from a total force of 245,000) in support of complex crisis-management operations, the *Bundeswehr* is struggling to maintain the 8,300 troops currently deployed overseas (Brune *et al.*, 2010: 2). However, the current reform process initiated by former Defence Minister Thomas de Maizière (2009–13) has suspended conscription (Pidd, 2011). This step will free vital funds for investment in deployability and allow an increase in the proportion of troops available for deployment to 10,000.

German defence reform has also been characterised by investment in C4ISTAR capabilities to facilitate networked operations and the projection of higher-intensity military force within and outside of Europe's geopolitical neighbourhood.[1] A number investment programmes in the area of C4ISTAR have been instigated since 2003, which have put in place some important foundations for a network-enabled *Bundeswehr*. The first key area of investment is the development of joint, networkable radio equipment in order to boost C2 capabilities and interoperability with EU and NATO partners (Adams and Ben-Ari, 2006: 55–8; Dyson, 2010: 52–3). Second, investment has cohered around the creation of a joint intelligence, surveillance, and reconnaissance network (ISR), including space-based reconnaissance systems and a variety of UAVs (Adams and Ben-Ari, 2006: 56–7; Flournoy and Smith, 2005: 91–3; Matzken, 2007: 73–4).

Despite these acquisitions, the *Bundeswehr* continues to lag behind its key NATO partners in networked capabilities. NEC is composed of three stages: an initial stage, where links between existing equipment are improved, a transitional phase where the integration of systems is enhanced, and, finally, a mature state in which all possible military equipment is synchronised (UK Ministry of Defence, 2004: 1). By 2007 British and French forces enjoyed an initial networked capability (Adams and Ben-Ari, 2006: 29–33; Dyson, 2010: 40–3; Flournoy and Smith, 2005: 91). A demonstrator exercise had been planned for late 2012 to test the *Bundeswehr*'s initial network-enabled capability, but in late 2011 this exercise was postponed indefinitely.[2] This postponement is a consequence of doctrinal disagreement between the individual services over the implications of networking for command and control.[3]

However, it is also the result of technical problems in interoperability between the command and control systems of the services as too little consideration was given to jointness and to interoperability in procurement.[4] The individual services of the *Bundeswehr* have developed their own C2 systems which are unable to properly communicate with each other.[5] Indeed, of the *Bundeswehr's* 70 procurement programmes in 2010 only six were network-ready, 11 had limited NCW capability, while the remaining programmes were not network-ready (Wiesner, 2011: 15). Despite the identification of these problems, as Wiesner (2011: 15) notes: 'the identification of these problems did not result in a refocusing of command-and-control systems projects. Quite the contrary, service-specific communication platforms ... continue to be funded'. In addition to problems in C4ISTAR acquisition, until recently Germany has continued to invest too heavily in capabilities more suitable for Cold War conflict scenarios than for expeditionary warfare (Dyson, 2011).

Furthermore, while Germany has been willing to use UAVs for reconnaissance purposes, it has shied away from the development and use of UCAVs. Germany is in the process of establishing a potential UCAV capability through the development of the EADS Barracuda medium-altitude long-range UAV in a joint procurement programme with Spain (Barracuda, 2013). Although the Barracuda does not currently include offensive capabilities it can be easily refitted to carry weapons. The 2009–13 CDU/CSU coalition had considered purchasing a number of Reaper UCAVs from the US in order to provide assistance to German special-forces missions against terrorist targets (Werkhaeuser, 2013). However, the CDU/CSU/Social Democrat (SPD) Grand Coalition elected at the September 2013 federal elections will slow – and may even block – the purchase of UCAVs. Under pressure from the SPD, the December 2013 Coalition Treaty has postponed the acquisition of UCAVs in order to enable a thorough consideration of the legal and ethical issues raised by the use of such capabilities.

Third, German emulation of the RMA has involved doctrinal development around the implications of networking for Command and Control and Effects-Based Thinking (Dyson, 2010: 47–56). Doctrinal development has also cohered around the development of Stabilisation doctrine (Noetzel and Schreer, 2009). In this regard German defence reforms represent a case of convergence with those of its closest European partners (Dyson, 2010: 28–60). However, this convergence in doctrine is also partial and delayed.

Despite the development of a consensus within the *Bundeswehr* that technology has not fundamentally changed the character of warfare, there remains a high-level of contestation over the implications of the RMA for Command and Control (C2). German thinking on NEC has taken the form of *Vernetzten Operationsführung* (Networked Operational Command, Doctrine *NetOpFü*).[6] However, significant contestation exists within the German military over utility of Effects-Based Thinking and of

NetOpFü more generally. Indeed, the 2009 Operation Common Shield, that formed an early test of the *Bundeswehr's* networking capabilities, identified the lack of a common position on *NetOpFü* between the services as a major problem (Heitmüller, 2009). Nevertheless, German thinking about the implications of networked capabilities for military doctrine remains something of an academic exercise, due to the delays in Germany's procurement of the systems necessary to attain an initial network-enabled capability.[7]

The *Bundeswehr* also suffers from significant deficits in its Stabilisation and COIN doctrine. In the UK and France, doctrine provides a detailed account of how to undertake Stabilisation and COIN operations at the tactical and operational levels (Alderson, 2010: 28–40; De Durand, 2010: 11–28). Crucially, British and French army doctrine has, since 1997, recognised that expeditionary crisis-management operations are likely to be characterised by conflict that can vary rapidly in intensity across the conflict spectrum (Dyson, 2010: 34–47). Yet, until it was revised in 2007, German Army doctrine (*Truppenführung von Landstreitkräften*) categorised fighting, peace-support, and humanitarian aid into distinct categories of conflict (Noetzel and Schreer, 2009: 16–22). Doctrine was updated in 2005 to include Guidelines for Operations against Irregular Forces; however, it did not properly integrate the kinetic and non-kinetic dimensions of military operations (Noetzel and Schreer, 2009: 16–22). As a consequence, the *Bundeswehr* lacks a comprehensive COIN doctrine suitable for the kind of operations it has encountered in the north of Afghanistan.[8]

In summary, the defence reforms of Britain, France, and Germany have been characterised by a significant level of convergence. This convergence has taken the form of a partial and selective emulation of the force structures, capabilities, and military doctrines associated with the RMA. Central to this partial emulation of the RMA is scepticism of the potential of technology to fundamentally transform the nature of conflict and recognition of the need to adapt technological advances to the needs of conflict scenarios which have the potential to vary quickly in intensity across the conflict spectrum. However, despite this convergence, the *Bundeswehr* remains an outlier case. Germany's defence reform process has been characterised by significant temporal delay, reducing its ability to undertake the kind of precision strike, networked operations that the British and French armed forces are capable of. The following section explains these patterns of convergence and divergence by analysing the theoretical insights of cultural and realist approaches to the sources of military change.

Theorising military emulation: the imperatives of military best practice

Cultural and Realist approaches form the dominant theoretical models in the literature on the sources of military change. Cultural approaches emphasise the centrality of social construction and norms in policy formation. Hence the literature on national 'strategic culture' places analytical weight on the role played by cognitive paradigms within societies and the key institutions of defence and security policy making in 'predisposing societies in general and political elites in particular to certain actions and policies over others' (Duffield, 1998: 27).

In their studies on German and French defence reform, Longhurst (2003) and Irondelle (2003) emphasize the role of 'path dependency'. The concept of 'path dependency' is rooted in historical institutionalism's assertion that institutions embody 'formal or informal procedures, routines, norms and conventions' and tool-kits of action which provide actors with ways of defining problems and 'logics of appropriate behavior' (Hall and Taylor, 1996: 938).

Hence according to cultural approaches, the timely emulation of the RMA in the UK and France is a consequence of British and French strategic culture/path dependency. The global role played by both states during the colonial era meant that the institutions of British and French defence and security policy, as well the broader populations of these states, were prepared for the post-Cold War operational environment – the challenges of expeditionary warfare of varying intensity (Irondelle, 2003; Thornton, 2003). German strategic culture, on the other hand, is determined by the legacy of the moral and military defeat of World War II. According to proponents of strategic culture (Dalgaard-Nielsen, 2006; Longhurst, 2004) these seismic historical events have fostered a deeply embedded and powerful culture of anti-militarism, reticence towards the use of high-intensity military force as an instrument of German foreign policy, and a particularly strong preference to work through multi-lateral institutions in the deployment of military force.

Culturalists, such as Farrell (2005: 450), also argue that norms of 'conventional warfare' exist not only at the national, but at the global level, 'which provide the basic template for military organisation'. These norms, Farrell argues, are embedded within transnational organisational fields and are disseminated through professional networks. They form the central means through which ideas about the appropriate model of military reform in the contemporary international system – i.e. the partial and selective emulation of the US-led RMA – are disseminated. Where domestic norms (organisational and strategic culture) dovetail with these transnational norms, their adoption will proceed unchallenged. When there is greater disjuncture between domestic norms and transnational norms the process of military change will be slower (Farrell, 2005).

However, Neorealism also provides a compelling explanation for military change. Neorealism argues that the international system is anarchic. The lack of an authority above the state to regulate international interaction action fosters a high level of insecurity as states can never be truly certain of the intentions of other states. Anarchy rewards those who conform to its logic with an increase in power and security, and punishes those who do not with decline and military defeat (Waltz, 1979: 89–93). Hence the anarchic logic of the international system drives the adoption of new military methods (Posen, 1984). The importance of military power to a state's relative power in the international system and its ability to survive in an uncertain, 'self-help' world, leads to three possible outcomes (which can be combined to varying degrees).

First, the continuation of existing practices (policy inertia), where states do not adapt to the imperatives of the security environment and retain existing military objectives and instruments. Such policy inertia will, over time, lead to a loss of power and influence as the state's military effectiveness declines. The second outcome is emulation, which constitutes the adoption of military 'best practice':

> The possibility that conflict will be conducted by force leads to competition in the arts and instruments of force … contending states imitate the military innovations contrived by the country of greatest capability and ingenuity … competition produces a tendency towards sameness of the competitors … and so weapons of major contenders and even their strategies begin to look much the same.
>
> (Waltz, 1979: 127)

According to Waltz (ibid.) the distribution of material capabilities in the international system (the balance of power) is the central driver of military emulation. States will adopt the military practices of the most powerful state in the international system, that, by virtue of its dominance in relative power, is worthy of emulation. However, as Resende-Santos (2007: 51, 58–61) notes, rather than emulating the state of greatest capability in the international system, the emulation of proven success in conflict is the quickest and most effective strategy that involves the least risk. The central factor driving military emulation is therefore 'competitive effectiveness': the extent to which a state's defence policy will allow it to maximise its power and influence and ensure its survival, autonomy, and prosperity (Resende-Santos, 2007: 63). Hence the key criterion in determining the attractiveness of emulation is proven success in great power war (Resende-Santos, 2007: 81–2), as dominance in material capabilities does not preclude faulty decision making in force structures, capabilities, and doctrine by the leading state in the international system.

The final possible outcome in defence reform is innovation: 'the discovery of new knowledge, invention of new practices or their recombination in new forms' (Resende-Santos, 2007: 72). Innovation offers the potential for the development of a significant advantage over other states. The advantages delivered by innovation are inevitably short-lived because of the rapid diffusion of military technology and practices, allowing other powers to copy successful innovation. Innovation is also associated with a high-level of risk because of the potentially catastrophic implications of poor strategic choices.

The likelihood of a state to choose a strategy of innovation or emulation is a consequence of the interaction between two key factors. The first factor is a state's resource-technological basis. The greater the capabilities of a state, the lower the risk that is associated with failed innovation (Resende-Santos, 2007: 73). The second factor is whether credible alliance options are available, which can provide an opportunity to bandwagon on the power and innovations of other states (ibid.: 88–92).

Europe's selective emulation of the RMA – and its timing – is a consequence of both these factors. The nature of the West European Great Powers as 'secondary states' has predisposed these states to a strategy of selective emulation of US military innovations. Military innovation in the post-Cold War era was characterised by a high level of risk, as the utility of new precision strike capabilities were unclear in the strategic flux of the early 1990s. Emulation has proceeded on the basis of the identification of military 'best practice' gathered through the observation of the experiences of Alliance partners with the RMA. Emulation has also been strongly informed by 'lessons-learned' processes of the militaries of the West European Great Powers (Dyson, 2012).[9] The institutional structures of 'lessons-learned' within the British, French, and German armed forces have undergone significant improvement during the post-Cold War era. These 'organisational capabilities' have provided important guidance on the implications of European states' operational experiences with the RMA, as well as the experience of other states in the international system, allowing a more objective identification of the imperatives of military 'best practice'.

The presence of Alliance options has also had an important impact on the timing of Europe's partial emulation of the RMA. The opportunity for the West European Great Powers to bandwagon on US capabilities through NATO during the 1990s provided a window of opportunity to assess the extent to which the capabilities and doctrines associated with the RMA were of utility to the post-Cold War operational environment. However, in the context of uni-polarity, European states have been unable to continue to bandwagon so heavily on US military innovation. In order to retain a measure of influence over the US they have sought to become more credible alliance partners by enhancing their interoperability through NATO C4ISR and force generation initiatives. Second, as the strategic interests of the US have begun to shift away from Europe, European states have also had to develop networked capabilities through the CSDP as a means with which to 'go it alone' when necessary. These strategic imperatives were forcefully highlighted by the Kosovo Conflict of 1999 that demonstrated to the Europeans the limitation of their reliance on US capabilities to provide security in their geopolitical neighbourhood. NATO's Operation Allied Force acted as a key catalyst for the acceleration of European investment in C4ISR, on a national as well as a European level.

However, Neorealism is unable to provide a full theoretical explanation for the patterns of European defence reform outlined in the first section of this chapter. While Neorealism provides a compelling framework to conceptualise the selective emulation of the RMA, it emerges as a more limited tool in understanding the precise timing of reform. As the following section demonstrates, Neoclassical Realism provides the strongest analytical leverage in understanding the temporally uneven process of European defence reform.

Neoclassical realism: the temporality of emulation

Neorealism posits that executive decision makers enjoy a high level of autonomy in translating threats from the international security environment into changes to military doctrine and structures and in overcoming domestic opposition and organisational politics. The theory argues that the ability of policy makers to drive change is particularly strong as the level of external threat increases and models of military best practice become clear, due to the potentially catastrophic implications of the loss of relative power that can result from faulty strategic choices (Posen, 1984). It also posits that states of relatively similar material capabilities, size, and population, such as Britain, France, and Germany, should exhibit isomorphic responses to the imperatives of the international security environment. The theory's analytical leverage is therefore found wanting in explaining the slow transmission belt between changes in the international security environment and changes to German military structures, capabilities, and doctrine, highlighted in the first section of this chapter.

However, these theoretical problems can be overcome through the use of Neoclassical Realist theory. Neoclassical Realism blends the insights of Neorealism and Classical Realism and in so doing, provides a mechanism with which to understand the role and nature of the domestic variables which act in an intervening manner to slow down or speed up convergence with the imperatives of military 'best practice'. In accordance with the fundamental premises of Neorealism, Neoclassical Realism argues that over the long term states will attempt to maximise international influence according to the constraints and opportunities presented by their own material power and the balance of power (Rose, 1998: 152). However, like Classical Realism, the approach is also interested in understanding the impact of domestic-level variables. It argues that 'state power'[10] is highly varied and acts as the main intervening unit-level variable that explains divergence, over the short to medium term, with the dictates of the international system (Rose, 1998: 152; Taliaferro *et al.*, 2009: 25).

Neoclassical Realism has received a significant level of attention in recent years and substantial contestation has emerged about which domestic variables are dominant in determining 'state power'. Neoclassical Realist scholars such as Schweller (2004: 169), Snyder (1991), Taliaferro (2006), Van Evera (1999), and Zakaria (1998: 38) place causal weight upon a wide number of variables which determine 'state power', including nationalism, ideology, culture, religion, ethnicity, and state-society relations. This section demonstrates that the timely British and French convergence with the imperatives of the international security environment and temporal delay in German emulation of the RMA is a product of the level of autonomy enjoyed by the political executive. Executive autonomy in defence policy is determined by a set of domestic material power relations: notably the institutional structure of the state (federal or unitary) and the linkages between policy systems (foreign, social, budgetary, and defence policy).

The majority of the literature on German defence policy posits that German strategic culture is an important determinant of the content and timing of changes to German defence policy (Berger, 1998; Dalgaard-Nielsen, 2006; Longhurst, 2004). According to this scholarship, the retention of conscription until 2010 and the reticence of German policy makers to sanction the development of a military capable of higher-intensity networked military operations are the legacy of Germany's past. However, a focus on the constraints presented by Germany's federal political system provides a more convincing explanation of Germany's slow process of reform. The institutional structure of the German state and linkages between policy subsystems of defence, budgetary, and social policy has narrowed the autonomy of the core executive to make far-reaching changes to the content of German defence policy and, by extension, to the institutional structures of German defence politics.

First, the Federal political system has created a high level of concern amongst German Defence Ministers with the politics of base closures. Military bases are often responsible for maintaining thousands of jobs and engage the interest of the Chancelleries and Economics Ministries of the German *Länder* (states) (Dyson, 2007). The frequent regional elections of the German federal system and power that individual *Länder* wield through the *Bundesrat* sensitises the core executive to *Länder* concerns. Defence Ministers were, therefore, very keen to avoid the abolition of conscription and the consequent process of force downsizing. This problem was compounded by the ambitions which were harboured by several Defence Ministers in the post-Cold War era (notably Volker Ruehe (1992–98), Rudolf Scharping (1998–2002), and Karl-Theodor zu Guttenberg (2008–10) of becoming Chancellor. Far-reaching base closures would incur the wrath of powerful regional politicians in their own political parties and were therefore to be avoided (Dyson, 2007). In contrast to Germany, in the UK and France, the unitary state provides significant windows of opportunity to push through unpopular large-scale base closures.

Second, the autonomy of the German core executive in the field of defence policy is also restricted by the way that the federal political system restricts the core executive's ability to mobilise the public in support of tackling contemporary security challenges. There is little doubt about the sensitivity of German public opinion to the use of the *Bundeswehr* for higher-intensity operations. However, the roots of German 'security culture' lie in a set of material forces.

As Hyde-Price (2007: 46–7) highlights, when a state is subject to particular systemic imperatives for a long period of time, they create 'channels' in which political activity settles. During the Cold War the principles of anti-militarism and multilateralism developed as a rational response to the security environment that German faced following the end of World War II. The development of an anti-militaristic security narrative and the embedding German defence and security policy within multilateral institutions like the UN and NATO, were central instruments in security Germany's re-emergence and rehabilitation into the community of Western nations. It was also a key mechanism for ensuring that West Germany was able to gain US, British, French, and Russian support for eventual German reunification (Cole, 2001: 6–12; Sutton, 2007: 91).

By the end of the Cold War, these imperatives and the security narratives which surrounded them had taken on a significant level of ideological rigidity and 'path dependency' within the institutions of German defence and security as well as within German society. However, these security narratives required significant reworking in the context of new post-Cold War security challenges which necessitated the deployment of high-intensity expeditionary military force.

For the British and French, the security challenges of the post-Cold War era were much more congruent with the military roles undertaken by these states as colonial powers and required less fundamental reshaping. However, the level of 'fit' between colonial and post-Cold War security challenges is not the only variable at play. It is also important to note that the British and French unitary political systems offer significant windows of opportunity to make unpopular decisions about expeditionary troop deployment and alter some of the central tenets of their defence policies. This high level of autonomy is well illustrated by the way that, following his election in 2007, former French President Nicholas Sarkozy was able to refashion ideology by employing the principles of Gaullism to justify French reintegration to NATO, arguing that reintegration to NATO was a 'break with the method, but not the principles of NATO' and that it would help to enhance French 'independence' (France24, 2009). This action was, of course, in contradiction to the principles of national strategic autonomy and suspicion of Anglo-Saxon security arrangements which characterise Gaullism.

In contrast to France and the UK, regular elections at the *Land* level create a complicated process of reshaping the narrative used to mobilise society on behalf of responding to the exigencies of the contemporary security environment.[11] Rather than the bold and assertive policy leadership of President Sarkozy, German policy makers have faced a domestic political context that incentivises a much less ambitious 'salami-slicing' approach to changing the role of the German military that undertakes frequent, but much more limited, changes to defence policy (Dyson, 2007: 69–71).

The connections between German defence, budgetary, and social policy have also acted to reduce the autonomy of the core executive in defence and slowed the pace of German reform, reducing the level of funds for investment in new capabilities, such as C4ISR. Until 2010 Germany's system of conscription was accompanied by an extensive system of conscientious objection that allowed between 80,000 and 130,000 young men to serve in the social system at a third of the cost of a professional in a similar position. During the 1990s and 2000s the presence of this large pool of cheap manpower led the Federal Finance Ministry to be highly reluctant to abolish conscription (Dyson, 2007). The abolition of conscription would have added to the fiscal burden at an inopportune time due to the imperative of fiscal transfers to the East of Germany to reconstruct the former DDR, while attempting to comply with the constraints of the Stability and Growth Pact of European Monetary Union. Only in 2010, once it became abundantly clear that conscripts were no longer of any use in contemporary military operations, did Germany abolish conscription.

While Britain had abolished conscription in 1957, France had an extensive system of conscription in place at the end of the Cold War. However, conscription was abolished in a dramatic manner in 1996 by French President Jacques Chirac (Irondelle, 2003: 157–87). This bold policy leadership by the President was facilitated by the fact that France did not have a widespread system of conscientious objection as well as the large window of opportunity to enact radical change that was afforded by a seven-year term as President.

These three factors have helped to create a dysfunctional model of civil-military relations in defence planning which have slowed the ability of the German military to identify and implement 'military best practice' and its emulation of the RMA. The incentive provided by low executive autonomy in defence to slow the pace of defence reform has led Defence Ministers to deliberately block processes of 'bottom-up' learning from operations and stunt the German lessons-learned process (Dyson, 2012). Hence Ministers have failed to enhance the intellectual dynamism of the *Bundeswehr*, not only on questions of force structures and the persistence of conscription, but also in areas such as military doctrine, where low executive autonomy has had a very detrimental impact (Dyson, 2012). 'Bottom-up' learning processes about the implications of military operations for military doctrine had the potential to upset the careful 'salami-tactic' of gradual changes to the role of the German armed forces. Hence civilian policy-leaders have played an active role in blocking the ability of the *Bundeswehr* to tackle any lessons from theatre that sit uncomfortably with perceived electoral imperatives, not least the development of an explicit COIN doctrine and thinking on network-centric warfare (Dyson, 2012: 39–44).

Furthermore, the persistence of Germany's 'culture of restraint' in defence and security policy has led policy makers to be highly resistant to the deployment of high-intensity manned precision strike capabilities, but also to use of unmanned precision strike capabilities such as UCAVs for the targeted killing of terrorist suspects. Hence, while the CDU/CSU has considered the acquisition of Reaper drones, the Social Democratic Party (SPD) is highly resistant to such remotely piloted precision strike capabilities.[12] The presence of the SPD in a Grand Coalition with the CDU/CSU in the 2013–17 electoral period will complicate the acquisition of these capabilities. Indeed, the December 2013 Coalition Treaty between the CDU/CSU and SPD postpones the purchase of UCAVs to allow for further consideration of the legal and ethical issues associated with their use (Vestring, 2013).

This situation stands in stark contrast to the UK, where high executive autonomy in defence means that the military is allocated a great deal of freedom to facilitate adherence with doctrinal best-practice and deploy military force with less public scrutiny (Dyson, 2012: 34–9). Civilian intervention in doctrinal development in the UK and France has not attempted to dictate the content of doctrine. Rather, civilian policy makers have attempted to establish institutional structures, such as the Development Concepts and Doctrine Centre in the UK and Joint Forces Centre for Concept Development, Doctrine and Experimentation in France which are explicitly tasked with promoting objective joint doctrinal development (Dyson, 2012: 34–9; Dyson, 2010: 153–7).

The dynamism of German military procurement has also been restricted by low executive autonomy and has contributed to a dysfunctional model of civil-military relations in defence planning. The federal political system with its frequent *Länder* elections and powerful regional politicians makes it very problematic to make radical changes to defence capability procurement and reform force postures.[13] Consequently within the core executive there is little incentive to reform the weapons acquisition process and enhance its dynamism, leading to limited civilian intervention in defence planning. While the *Bundestag's* Budgetary Committee has the right to sanction procurement projects over 25 million Euros, it has little influence over the nature of projects which are presented to it and is more concerned with ensuring that projects benefit Germany industry than with their military utility.[14] The poor level of civilian oversight over military procurement was highlighted most recently by the disastrous procurement of the EUROHAWK surveillance drone that was developed jointly by EADS and Northrop Grumman. Despite investing over 680 million euros in the 12-year project the Eurohawk was cancelled in 2013 due to the lack of an anti-air collision system which would not allow it to operate in European airspace. The UK's high levels of executive autonomy are not matched by an efficient system of defence capability procurement, due to poor civilian management of decision making processes surrounding defence acquisition (Gray, 2009). However, in France, high executive autonomy and a highly centralised and civilian-led procurement process has fostered a far more efficient and dynamic system of weapons and equipment acquisition than exists in Germany (Gray, 2009).

Conclusions: the European imperative

For Europe to emerge as an autonomous military actor and retain influence within the US the capacity to undertake stand-off precision strike operations will be essential. As this chapter highlights, European states have undertaken a partial and selective emulation of the force structures, capabilities, and doctrines associated with the Revolution in Military Affairs. This isomorphism accords with the insights of Neorealism: that states are motivated by the imperative of attaining security in a competitive international security environment, However, Neorealism is of only limited utility in explaining the European defence reform, particularly the slow pace of German military reform. Instead Neoclassical Realism, with its emphasis on the intervening impact of the autonomy of the core executive emerges as the theoretical framework of greatest utility in understanding the pace of the convergence of European states around the selective emulation of the RMA.

While German military isomorphism is beginning to pick up speed, a further set of challenges now face European nations in defence. First, European nations face the challenge of the acceleration of the shift in US priorities away from Europe towards Asia in the context of the rise of Chinese power in the Western Pacific. The ramifications for the West European Great Powers are clear: take up a greater burden of European security or allow a security gap to develop in Europe and its geopolitical neighbourhood. The ability of European nations to carry a more substantial level of the security burden is, however, circumscribed by the second major challenge that they face: financial austerity and consequent defence cuts.

It is, therefore, vital that European nations begin to coordinate their defence cuts in the light of the potential for further cooperative force and capability pooling and sharing arrangements under the auspices of the EU's Common Security and Defence Policy's Ghent Framework and the European Defence Agency. However, a number of factors, which reflect the systemic and domestic insights of Neoclassical Realism, serve to undermine Europe's ability to consolidate its defence industry and pool and share capabilities.

First, European cooperation in defence is limited by the anarchic international system. The concept of the Alliance Security Dilemma was developed by Snyder (1984) and posits that whilst the balance of power may push states, like the West European Great Powers, toward convergence in military affairs and foster cooperation at the regional level, states cannot be fully certain that others in an alliance will not renege on their promises. This uncertainty about the commitment of Alliance partners fosters a high level of reticence to pool and share capabilities in a manner that may restrict national strategic autonomy.[15]

The second factor is the embeddedness of defence postures. As in Germany, ideological narratives have been employed in Britain to justify the particularly close relationship that emerged with the US and other English-speaking nations after the end of World War II. Systemic imperatives, not least the 'Asia pivot' of the US, point to the necessity of greater cooperation between the UK and its European partners in defence. However, the leader of the Conservative Party, Prime Minister David Cameron is constrained by the virulent Euroscepticism that has emerged within the Conservative Party. The Eurosceptic ideology associated with crafting the UK's strong Atlanticism and attachment to the Commonwealth begins to take on a path dependency that is difficult to alter, at least under the conditions of a coalition government, where executive autonomy is narrowed by the capacity of backbenchers to inflict defeat on the government and by the electoral threat presented by the UK Independence Party (UKIP) under a hung parliament. However, as Neoclassical Realism predicts, the West European Great Powers will eventually have to converge with systemic imperatives and pursue more – if necessarily limited – cooperation through CSDP. The alternative is a slow decline in their international influence and growing insecurity within Europe and its geopolitical neighbourhood.

Notes

1 Interview, Section for Concept Development and Experimentation, *Bundeswehr Transformation Centre*, Strausberg, Germany, 15 June 2012.
2 Interview, *Bundeswehr Transformation Centre*, Strausberg, Germany, 19 December 2011.
3 Interview, Section for Concept Development and Experimentation, *Bundeswehr Transformation Centre*, Strausberg, Germany, 15 June 2012.
4 Ibid.
5 Interview, *Bundeswehr Transformation Centre*, Strausberg, Germany, 19 December 2011; Interview, Section for Concept Development and Experimentation, *Bundeswehr Transformation Centre*, Strausberg, Germany, 15 June 2012.
6 Interviews, *Bundeswehr Leadership Academy*, Hamburg, Germany, 8 December 2010; interview, Section for Concept Development and Experimentation, *Bundeswehr Transformation Centre*, Strausberg, Germany, 15 June 2012.
7 Interview, *Bundeswehr Transformation Centre*, Strausberg, Germany, 19 December 2011.
8 Interviews, Interviews, Division 1 (Army Development), Division 2 (Training) and Department for Coordination of Lessons-Learned, Heeresamt, Cologne, 18 May 2011.

9 It is important to note that the transmission belt linking changes in the international security environment to military change is an imperfect one. While the challenges of the operational environment is the key independent variable driving change, bureaucratic politics between the single services and deficits in the autonomy of the core executive in defence policy have complicated the process of the identification of military 'best practice' in Britain, France, and, in particular, Germany (see following section). On the impact of executive autonomy and bureaucratic politics on the temporality of British and French defence reform, see Dyson (2010).
10 State power refers to 'the impact of the strength of a country's state apparatus and its relations to the surrounding society upon ability of the state to mobilise and extract resources from society'. See Rose (1998: 152).
11 Interview, Defence and Security Policy Division, Federal Foreign Ministry, Berlin, Germany, 9 March 2012.
12 Interview, Rainer Arnold, MP, SPD Spokesperson on Defence Policy, German Bundestag, 3 July 2013.
13 Interviews, *Bundeswehr Transformation Centre*, Strausberg, Germany, 26 November 2009.
14 Interview, Division for International Armaments, Political Department, German Defence Ministry, Berlin, 29 May 2013.
15 Interview, Chancellor's Office, Berlin, Germany, 1 August 2012; Interviews, FDP *Bundestagsfraktion*, German Federal Parliament, 8 November 2012; 11 September 2012.

References

Adams, G. and Ben-Ari, G. (2006) *Transforming European Militaries: Coalition Operations and the Technology Gap*, Abingdon: Routledge.

Alderson, A. (2010) 'Britain', in Rid, T. and Keaney, T. (eds) *Understanding Counterinsurgency: Coalition Operations and the Technology Gap*, Abingdon: Routledge, pp. 28–46.

Baker, J. *et al.* (2013) *Guardian* (11 June 2013), 'End the Secrecy over UK Drones', www.guardian.co.uk/world/2013/jun/11/end-secrecy-uk-drones (accessed 7 July 2013).

Barracuda (2013) '10 Years of Development of Unmanned Flight Technology', www.airbus-group.com/airbusgroup/int/en/news/press.20130618_cassidian_barracuda.html (18 June 2013) (accessed 13 January 2014).

Berger, T. (1998) *Cultures of Anti-Militarism: National Security Cultures in Germany and Japan*, Baltimore: Johns Hopkins University Press.

Brune, S.-C., Cameron, A., Maulney, J.-P., and Terlikowski, M. (2010) *Restructuring Europe's Armed Forces in Times of Austerity*, Stiftung Wissenschaft und Politik Comments 28.

Cole, A. (2001) *Franco-German Relations*, Harlow: Pearson.

Dahl, E. (2002) 'Network Centric Warfare and the Death of Operational Art', *Defence Studies*, 2, 1: 1–24.

Dalgaard-Nielsen, A. (2006) *Germany, Pacifism and Peace-Enforcement*, Manchester: Manchester University Press.

De Durand, E. (2010) 'France', in Rid, T. and Keaney, T. (eds) *Understanding Counterinsurgency: Doctrine, Operations and Challenges*, Abingdon: Routledge, pp. 11–28.

Duffield, J. (1998) *World Power Forsaken: Political Culture, International Institutions and German Security Policy after Unification*, Stanford: Stanford University Press.

DWP (2013) *French Defence White Paper on Defence and National Security: Twelve Key Points,* Paris: Odilie Jacob.

Dyson, T. (2007) *The Politics of German Defence and Security: Policy Leadership and Military Reform in the post-Cold War Era,* New York: Berghahn.

Dyson, T. (2008) 'Between International Structure and Executive Autonomy: Convergence and Divergence in post-Cold War British, French and German Military Reforms', *Security Studies,* 17, 4: 725–74.

Dyson, T. (2010) *Neoclassical Realism and Defence Reform in Post-Cold War Europe,* Basingstoke: Palgrave Macmillan.

Dyson, T. (2011) 'Managing Convergence: German Military Doctrine and Capabilities in the 21st Century', *Defence Studies,* 11, 2: 244–70.

Dyson, T. (2012) 'Organising for Counterinsurgency: The Adaptation of British and German Military Doctrine in Afghanistan', *Contemporary Security Policy,* 33, 1: 27–58.

Farrell, T. (2005) 'World Culture and Military Power', *Security Studies,* 14, 3: 448–88.

Farrell, T. (2008) 'The Dynamics of British Military Transformation', *International Affairs,* 84, 4: 777–807.

Flournoy, M. and Smith, J. (2005) *European Defense Integration: Bridging the Gap between Strategy and Capabilities,* Washington, D.C.: Center for Strategic and International Studies.

France24 (2009) www.france24.com/en/20090311-france-nato-commandstructure-wider-role-troops-charles-de-gaulle (accessed 29 November 2012).

Gray, B. (2009) *Review of Acquisition for the Secretary of State for Defence,* London: HMSO.

Hall, P. and Taylor, R. (1996) 'Political Science and the Three New Institutionalisms', *Political Studies,* 44, 5: 936–57.

Hauser, B. (2007) 'The Cultural Revolution in Counter-Insurgency', *Journal of Strategic Studies,* 30, 1: 153–71.

Heitmüller, U. (2009) 'Common Shield: Die Bundeswehr Testet Neue Technologien', *Die Welt* (13 February 2009).

Hyde-Price, A. (2007) *European Security in the 21st Century: The Challenge of Multipolarity,* Abingdon: Routledge.

Irondelle, B. (2003) 'Civil-Military Relations and the End of Conscription in France', *Security Studies,* 12, 3: 157–87.

Longhurst, K. (2003) 'Why Aren't the Germans Debating the Draft? Path Dependency and the Persistence of Conscription', *German Politics,* 12, 2: 147–65.

Longhurst, K. (2004) *Germany and the Use of Force: The Evolution of German Security Policy 1989–2003,* Manchester: Manchester University Press.

Marx, B. (2013) *Deutsche Welle* (31 May 2013), 'Growing Criticism over Eurohawk Debacle', www.dw.de/growing-criticism-over-euro-hawk-debacle/a-16852393 (accessed 13 January 2014).

Matzken, M. (2007) 'German Army Stabilisation Forces', *Doctrine,* 12: 71–4.

Noetzel, T. and Schreer, B. (2009) 'Missing Links: The Evolution of German Counterinsurgency Thinking', *RUSI Journal,* 154, 1: 16–22.

Pidd, H. (2011) *Guardian* (16 March 2011), 'Marching Orders for Conscription in Germany, but what will Take its Place?', www.theguardian.com/world/2011/mar/16/conscription-germany-army (accessed 13 January 2014).

Posen, B. (1984) *The Sources of Military Doctrine: France, Britain and Germany between the Wars,* Ithaca: Cornell University Press.

Resende-Santos, J. (2007) *Neorealism, States and the Modern Mass Army*, Cambridge: Cambridge University Press.

Reynolds, K. (2006) 'Building the Future Force: Getting Defence Transformation Right', *Contemporary Security Policy*, 27, 3: 435–71.

Rose, G. (1998) 'Neoclassical Realism and Theories of Foreign Policy', *World Politics* 51, 3: 144–72.

Schweller, R. (2004) 'Unanswered Threats: A Neoclassical Realist Theory of Underbalancing', *International Security*, 29, 2: 159–201.

Snyder, G. (1984) 'The Security Dilemma in Alliance Politics', *World Politics*, 36, 4: 461–95.

Snyder, J. (1991) *Myths of Empire: Domestic Politics and International Ambition*, Ithaca: New York.

Sutton, M. (2007) *France and the Construction of Europe 1994–2007: The Geopolitical Imperative*, Oxford: Berghahn.

Taliaferro, J. (2006) 'State Building for Future Wars: Neoclassical Realism Extractive State', *Security Studies*, 15, 3: 464–95.

Taliaferro, J., Lobell, S., and Ripsman, N. (2009) 'Introduction: Neoclassical Realism, The State and Foreign Policy', in Taliaferro, J., Lobell, S., and Ripsman, N. (eds) *Neoclassical Realism, The State and Foreign Policy*, Cambridge: Cambridge University Press, pp. 1–42.

Thornton, R. (2003) 'A Welcome Revolution? The British Army and the Changes of the Strategic Defence Review', *Defence Studies*, 3, 3: 38–62.

UK Ministry of Defence (2004) 'Future Capabilities, Factsheet 1. The Policy Baseline: Why We Need to Change', UK Ministry of Defence.

Van Evera, S. (1999) *Causes of War: Power and the Roots of Conflict*, Ithaca, NY: Cornell University Press.

Vestring, B. (2013) 'Coalition Cop-Out on Drones', German Council on Foreign Relations, https://ip-journal.dgap.org/en/blog/berlin-observer/coalition-cop-out-drones (accessed 13 January 2014).

Waltz, K. (1979) *Theory of International Politics*, Reading, MA: Addison-Wesley.

Werkhäuser, N. (2013) *Deutsche Welle* (5 May 2013), 'Drones for the German Military Stir Controversy', www.dw.de/drones-for-the-german-military-stir-controversy/a-16786841 (accessed 7 July 2013).

Wiesner, I. (2011) 'The Paradoxical Ally. Germany's Decoupling Strategy in the Adoption of Network Centric Warfare', paper presented at the 2011 International Studies Association Annual Convention, Montreal, March.

Zakaria, F. (1998) *From Wealth to Power: The Unusual Origins of America's World Role*, Princeton: Princeton University Press.

3 The global diffusion of unmanned aerial vehicles (UAVs), or 'drones'

Ulrike Esther Franke

Introduction

The proliferation of unmanned aerial vehicles (UAVs, or 'drones' in popular parlance) has dramatically increased over the last few years. Today between 56 and 78 countries hold unmanned aerial military capabilities. This chapter discusses current proliferation trends and provides an overview of the most important stakeholders in the global UAV trade. It will be argued that UAVs do not only proliferate because they have proven their military worth. Rather, the view has emerged that modern militaries need unmanned capabilities in order to appear capable and modern – UAVs are becoming 'must-haves' for militaries around the world. Efforts to ban UAVs or to increase their control through existing and new arms control instruments are hence likely to encounter resistance.

The popularity of UAVs has dramatically increased over the last decade. While in 2000 only 17 states held UAVs, today approximately 70 countries have unmanned military capabilities and their number is rising steadily (see Table 3.1). UAVs are being bought and sold internationally, and many countries invest in indigenous drone research and development (R&D) programmes. In 2011 a study estimated that there were 680 different UAV programmes worldwide (Wilson 2011).

This chapter begins by giving a short overview of UAV development and then goes on to discuss the diffusion of military UAVs since the year 2000. This date has been chosen because of the culmination of UAV developments in the early 2000s, such as the use of (unarmed) Predator UAVs for the tracking of insurgents in Afghanistan (2000), the first launch of a Hellfire missile from a Predator (2001), and the first victim of a missile fired by a UAV in 2002. The chapter analyses current proliferation trends, presents the most important UAV exporting nations, and discusses China's attempts to establish itself as an important stakeholder in the UAV market. Some indigenous R&D programmes and international joint ventures will be reviewed as examples.

It will be argued that the current global proliferation of UAVs cannot solely be explained by the neo-realist theory of military emulation. Rather,

Table 3.1 Current UAV users (data as of September 2013)

Known holders of unarmed UAVs (56)	South Africa
Algeria	Spain
Argentina	Sri Lanka
Australia	Sweden
Azerbaijan	Switzerland
Belgium	Taiwan
Botswana	Thailand
Brazil	Turkey
Burundi	Uganda
Canada	UAE
Chile	UK
China	US
Columbia	
Czech Republic	**Assumed holders unarmed UVAs (22)**
Denmark	Angola
Ecuador	Armenia
Egypt	Belarus
Estonia	Bulgaria
Finland	Croatia
France	Cyprus
Georgia	Ethiopia
Germany	Indonesia
Greece	Kazakhstan
Hungary	Kenya
India	Libya
Iran	Nigeria
Iraq	North Korea
Ireland	Panama
Israel	Philippines
Italy	Peru
Japan	Serbia
Jordan	Slovakia
Lebanon	Syria
Malaysia	Sudan
Mexico	Uruguay
Morocco	Venezuela
Netherlands	**Among these: Known holders of armed UVAs (3)**
New Zealand	
Norway	Israel
Pakistan	US
Poland	UK
South Korea	**Suspected holders of armed UAVs(s)**
Romania	China
Russia	Iran
Singapore	

the perception that modern militaries need unmanned capabilities in order to be accepted internationally, that UAVs are a 'must have' item for any modern military, is playing a crucial role in promoting the procurement of UAVs around the world. Efforts to ban UAVs or to increase their control through existing and new arms control instruments are hence likely to encounter resistance.

Unmanned Aerial Vehicles, a definition

There is no commonly accepted UAV definition. Until recently, NATO's Standardization Agency (2008: 2-U-2) defined a UAV as:

> A powered, aerial vehicle that does not carry a human operator, uses aerodynamic forces to provide vehicle lift, can fly autonomously or be piloted remotely, can be expendable or recoverable, and can carry a lethal or non-lethal payload. Ballistic or semi-ballistic vehicles, cruise missiles, and artillery projectiles are not considered unmanned aerial vehicles.

This definition was however abandoned in 2011 in favour of the less specific Unmanned Aircraft System (UAS): 'A system whose components include the unmanned aircraft, the supporting network and all equipment and personnel necessary to control the unmanned aircraft' (NATO Standardization Agency, 2013: 2-U-2).

For the purpose of this chapter and following common practice, a UAV is defined as:

> An airborne vehicle which does not carry a human operator, which may be piloted remotely, follow a pre-programmed flight path, fly autonomously, or a combination of all three. It is designed to be recoverable and carries a lethal or non-lethal payload. Non-recoverable vehicles and projectiles such as ballistic vehicles, cruise missiles, and artillery projectiles are not considered UAVs.

Note that this definition does not prescribe a minimum or maximum size of UAVs; current UAV types range from insect-sized models to aircraft of the size of commercial airliners. Neither is there a minimum or limit on take-off weight or payload. Also, the definition does not specify the functions for which UAVs can be used.

What's in a name?

There are several names used to describe Unmanned Aerial Vehicles. 'UAV', 'RPV'/'RPAS' (Remotely Piloted Vehicle/Remotely Piloted Air System), 'UAS' (Unmanned/Uninhabited Aerial System), and 'drone' are the most common. Historically, they used to describe different types of unmanned flying vehicles;[1] today the terms are largely employed synonymously. They do however carry different connotations.

The term drone is most widely used in the media and popular literature. Favoured by anti-UAV activists it has become increasingly unpopular with UAV manufacturers and the military. At the 2013 drone industry convention in Washington D.C. – much to the amusement of the attending journalists – the internet password was 'DontSayDrones' (Wolfgang 2013). The concern is that the term drone could evoke negative associations, of science fiction-like autonomous killer robots. UAV manufacturers and government or military officials aiming at emphasising the 'man-in-the-loop' prefer to use the term Remotely Piloted Vehicle. The UK Ministry of Defence advises 'when talking to the media' the use of the terms Remotely Piloted Vehicle (for the vehicle) and Remotely Piloted Air System (to describe 'the entirety of that which it takes to deliver the overall capability') (UK MoD, 2011: 2–1/2–2). UAV is presumably the most neutral expression. Ultimately, it should be noted that rather than referring to technical details, the decision regarding which term to employ is dependent on the author's view of UAVs and the target audience.

UAVs' early beginnings

The idea of using unmanned flying vehicles for military purposes is arguably as old as warfare itself, or dates back at least to manned aviation's infancy (Spinetta, 2010). Some authors trace the roots of UAV technology back a thousand years to Chinese troops' use of small lantern balloons for military signalling (Trautermann, 2006). Others consider the advent of hot-air balloons in the late sixteenth century as first step towards the developments of today's UAVs. In this context, the Austrian attack on Venice in August 1849 is often cited in which unmanned balloons carrying explosives were used (McDaid and Strong, 2003). Early surveillance UAVs such as the Ryan Firefly and Lightning Bug which were based on target drones – drones used for air defence training – were used by the US Air Force in Vietnam and flew some 3,400 sorties during the war (Garamone, 2002). Furthermore, UAVs were used by Israel in the 1982 Lebanon War and by the US in the First Gulf War.

Despite these early developments it was not until the late 1990s/early 2000s that modern UAV use began to gain momentum. Only the technological advances of the end of the last millennium – smaller, lighter, and cheaper camera technology; improved data storage and transmission; GPS – made modern UAV technology possible. The second major impetus for UAV development and in particular the development of armed UAVs was the 9/11 terrorist attacks on the World Trade Centre and the Pentagon. Not only did 9/11 and the ensuing military engagements result in increased defence spending, thereby facilitating the advent of a new military technology. It also led to UAVs being promoted as the ideal-type weapon to fight terrorists. UAVs' ability to secretly monitor persons of interest over prolonged periods of time made them attractive weapons for counter-terrorism. The 9/11 attacks also reduced the US reticence about carrying out targeted killings abroad (Mazzetti, 2013: 98). This development eventually paved the way for the extensive arming of UAVs and their use for targeted killings. In 2001 a US Predator UAV fired a Hellfire Missile for the first time in combat, in 2002 it claimed its first victim. By 2009, US drone strikes had, according to then-CIA Director Leon Panetta become 'the only game in town in terms of trying to disrupt the al-Qaida leadership' (Panetta 2009).

Global UAV proliferation since 2000 and current situation[2]

In *The Military Balance* of 2000, the International Institute for Strategic Studies lists only 17 countries holding military UAVs (IISS, 2000). The numbers of UAVs in the military arsenals of these 17 countries were mostly single-digit, with the notable exception of the US and Israel. No Latin American country had UAVs, nor did any Sub-Saharan African state besides South Africa. Eight years later, the number of UAV users had risen to 40 and UAVs were spread over all continents (IISS, 2008). Today, there are between 56 and 78 states which hold military UAVs (see Table 3.1). There are, however, great differences with regard to the sophistication and dimension of the respective countries' unmanned arsenals. Many countries only hold some small unarmed UAVs which are used by ground forces for close reconnaissance. Only about half of the UAV users have at their disposal the larger and more powerful MALE UAVs (Medium Altitude Long Endurance). To this date, comparatively few countries hold armed UAVs, namely the US, the UK, and Israel, as well as in all likelihood China and Iran. This situation is however about to change with several countries trying to procure or develop their own armed UAVs.

Considering these numbers, it is unsurprising that according to a recent market analysis, the unmanned sector has been the 'most dynamic growth sector of the world aerospace industry this decade' (Teal Group Corporation, 2013). The authors of the study predict that UAV spending will more than double over the next decade – 'from current worldwide UAV expenditures of $5.2 billion annually to $11.6 billion, totalling just over $89 billion in the next ten years' (ibid.).

A state resolved to procure UAV technology has two courses of action: buying off-the-shelf systems from allies, or developing and producing its own indigenous systems. A third, in-between option is to team up with an experienced partner and to manufacture a system on the basis of an existing one.

The decision for one of these options is influenced by a number of factors, such as what type of UAV and how many aircraft the state wants to procure, how urgently the military capability is needed, and how easy or difficult it is for the state in question to buy military equipment internationally. So far, military UAVs have been bought off-the-shelf when they were required for immediate use – as for instance for deployment to Afghanistan – or by countries wanting to get some first experience with the technology before deciding on whether to pursue it further. Furthermore, as Ted Harshberger, director of RAND Project Air Force (2012), points out:

> it is *very difficult* to build long-endurance, highly automated, multi-role unmanned systems of the sort often purchased by the United States and its allies. [...] Only a few companies – the major aerospace contractors in the United States and Europe, mainly – can produce these systems.

Indeed, for the moment only few countries have the expertise to produce technically sophisticated UAVs such as the larger MALE and HALE (Medium/High Altitude Long Endurance) models and armed models. Accordingly, these models are among the most sought-after types on the international market. On the other hand, 'it is *extremely easy* to produce modest-endurance, partially automated, single-purpose unmanned systems' (ibid.). Accordingly, the majority of indigenously produced UAVs are smaller, less sophisticated UAVs. UAVs have furthermore been produced locally by states unable to buy them abroad, for instance because of existing embargoes.

While for the moment two countries – the US and Israel – are frontrunners in the UAV market, this situation is poised to change with new actors such as China pushing into the market. Technological development is fast and many indigenous programmes are in the process of yielding promising results.

Indigenous UAV programmes

Most states with an unmanned military capability have some indigenous UAV types in their military arsenal. Particularly smaller, tactical, UAVs are often produced locally. An increasing number of states are also investing in more sophisticated UAV programmes.

All European armed forces employ domestically produced tactical UAVs, such as the Austrian Camcopter, the British Phoenix, the Czech Sojka III, the French DRAC, the German ALADIN, the Greek Pegasus, or the Italian Mirach 26. Because of urgent procurement needs, France, Germany, Italy, and the UK decided to import (or lease) foreign-built MALE UAVs as interim solutions for use in Afghanistan. In the long term, however, European defence companies plan to develop indigenous MALE (combat) UAVs. Several prototypes and UAV demonstrators have been developed to this end, such as the EADS Barracuda and Talarion/European UAS, the BAE Telemos and Taranis and the pan-European nEUROn.[3] In 2012 the UK and France agreed to share technical expertise gained from their respective UAV programmes to find out whether a joint Anglo-French combat UAV development/procurement programme would be viable (de Larrinaga, 2013a). For the moment, these projects are still in the development phase.

Outside of NATO, many states are investing in indigenous UAV programmes. Countries as diverse as Argentina, Armenia, Chile, China, Ecuador, Iran, Jordan, Malaysia, Russia, South Africa, Taiwan, and Venezuela currently use or develop their own indigenous UAVs.

Russia

Russia has a long history of indigenous UAV development as the Soviet Union had an advanced UAV programme. After the end of the Cold War, however, Russian UAV development stagnated and Russia missed out on the UAV development of the last decade (Axe, 2012). Most UAVs used by the Russian armed forces in the following years were Israeli-built. The Russian government appears now determined to catch up; in 2012 Russian President Vladimir Putin proclaimed that Russia would invest US$12 billion in indigenous UAV programmes by 2020 (Taylor, 2012). Russian manufacturers now reportedly aim at developing the full range of UAVs, including combat UAVs (Bell, 2012).

Iran

According to (most likely grossly exaggerated) Iranian government statements, Iran produces stealthy, long-range, and armed UAVs and has been mass-producing several UAV types (*Jane's Defence Weekly* 2007). Little reliable information is available on the extent of Iran's UAV programme. It seems that Iran is investing heavily in UAV research and that its UAV designs are increasingly sophisticated. It appears that Iran is focusing in particular on long-range UAVs, probably in order to be able to reach Israel (Binnie, 2012).

Iran's domestic UAV programme is in part born out of necessity as Iran cannot easily procure military equipment on the international market. Iranian UAV use dates back to the 1980–88 conflict with Iraq, when Iran used UAVs for tactical photo reconnaissance (ibid.). Iranian-made UAVs have made appearances in several conflicts. Hezbollah uses small unarmed Iranian-made UAVs (Sloggett, 2012). A *Global Post* investigation suggests that Syria's President Bashar al-Assad is using Iranian UAVs (Macleod and Flamand, 2012). Furthermore, Venezuela announced in 2012 that it had built a UAV with Iranian assistance (Paraszczuk, 2012).

China

China has been producing indigenous UAV for several decades. It started with reverse engineering both Soviet and US UAVs in the 1960s (Hsu, 2013: 3). Since the 1980s, China has been producing indigenous UAVs. For the moment, Chinese UAV production and development predominantly centres on small, tactical platforms, this situation is however changing. China appears to be investing heavily in MALE and combat UAVs – for domestic use as well as export.

At the 2012 Zhuhai Air Show, China revealed two MALE UAVs which could be armed; the Wing Loong (Chengdu Aircraft Design) and the CH-4 (China Aerospace Science and Technology Corp.). The Wing Loong can reportedly be equipped with guided missiles, miniature guided bombs, or laser-guided bombs. The CH-4 can drop laser-guided air-to-ground bombs (Minnick, 2012). So far, these UAVs have not been used in combat.

In recent years it has become clear that China is actively trying to position itself as key stakeholder in the global UAV trade. The growing number of exhibitors and the sophistication of the equipment and technology displayed at Chinese air shows demonstrate China's interest in marketing to other countries (Hsu, 2013: 15). Furthermore, there are indications that China is forming joint ventures with other states to produce UAVs. Pakistan recently presented a UAV at the International Defence Exhibition and Seminar (IDEAS) in Karachi whose design bore close resemblance to a Chinese UAV (Gordon and Hardy, 2012).

American UAV manufacturers are therefore worried that they are falling behind Chinese firms regarding the exports of strategic-level and in particular armed UAVs. A 2012 Department of Defense report warns: 'China might easily match or outpace U.S. spending on unmanned systems, rapidly close the technology gaps and become a formidable global competitor in unmanned systems' (Defense Science Board, 2012: 71). A Northrop Grumman employee has been quoted in a press article, 'if [the US] does not change its strict export restrictions, in five years' time we will be wondering how we could have gambled away our leading position so carelessly' (Lorenz *et al.*, 2011).

Indeed, Chinese firms seem to actively take advantage of the US reticence to export armed UAVs: 'The United States doesn't export many attack drones, so we're taking advantage of that hole in the market', a representative of the Chengdu Aircraft Design and Research Institute is quoted by the *Washington Post* (Wan and Finn, 2011). According to Chinese firms, some armed UAVs have already been exported to undisclosed recipients (Hewson, 2013). In the long run 'China's continued interest and progression in strategic-level UAVs appear poised to position China as a leader in the high-end UAV market' (Hsu, 2013: 14).

UAV export

Notwithstanding these various efforts, for the moment, the global UAV market is primarily dominated by two actors, Israel and the US, and their respective UAV manufacturers, most importantly Northrop Grumman, General Atomics, AeroVironment, IAI, and Elbit. The two states are so far the only two countries confirmed to have exported strategic-level UAVs (Hsu, 2013: 15).

Israel

Defence exports are an important element of Israel's economy and the sale of UAVs now accounts for nearly 10 per cent of the country's total defence export industry (Frost and Sullivan, 2013). Israeli UAVs are exported globally (with the notable exception of most Arab countries), currently, the top purchaser of these exports by spending is the UK; strong UAV trading relationships also exist with India and Brazil (ibid.). Israel primarily exports the IAI Heron (Heron I/Heron TP) and Elbit Hermes (Hermes 450/Hermes 900) MALE UAV series.[4] Heron UAVs have among others been exported to Australia, Brazil, Canada, Ecuador, Germany, India, and Turkey. Hermes aircraft are or will soon be used by the armed forces of Brazil, Chile, Croatia, Mexico, Singapore, and the UK. For the moment, these exported UAVs are all unarmed, but the Heron TP can be fitted with bombs and missiles if required.

United States

US UAV manufacturers have developed several UAVs which are of interest to foreign customers, most importantly the tactical AeroVironment Raven, the notorious General Atomics Predator and Reaper MALE UAVs, and the world's only HALE UAV, Northrop Grumman's Global Hawk.

The Raven UAV is, according to its manufacturer, 'the most widely used unmanned aircraft system in the world today' (AeroVironment, 2013). As the US Army alone uses several thousand of the hand-held, tactical UAVs, this statement is certainly true with regard to the numbers used. Raven UAVs have been exported to, amongst others, Uganda, Burundi, Lebanon, Norway, Spain, and the Czech Republic.

General Atomics' Predator and Reaper UAVs are unquestionably the best-known UAV models. Both types can be equipped with missiles or bombs but they can equally be used for surveillance missions. Predator and Reaper UAVs have the potential to become export hits as interest is strong in these well tested UAVs. So far, however, the US has been reluctant to export these models; currently, only Italy, France, and the UK fly Predator and Reaper UAVs. The UK is the only state to operate the armed version of the aircraft even though several US allies have expressed strong interest in armed Predators/Reapers. France acquired two unarmed Reapers in December 2013 with some 10 more vehicles to follow (de Larrinaga, 2013b). Algeria, Germany, the Netherlands, Turkey, and the United Arab Emirates have also expressed interest in the aircraft and are currently in negotiation with US authorities.

Northrop Grumman's Global Hawk is also being eyed by several states. NATO will procure five of the aircraft of the size of a commercial airliner for the alliance's long-awaited Alliance Ground Surveillance system (Holland, 2012). Canada, Japan, and South Korea are discussing the procurement of the aircraft.

Other exporters

Even though US and Israeli firms currently dominate the UAV market, nevertheless, some smaller players have succeeded in carving out a niche for their product. The Austrian-built Schiebel Camcopter S-100 is such a product. A vertical take-off UAV, it is particularly sought-after as it is able to take off from and land on ships, making it attractive for naval forces (Williams, 2011). The Camcopter has undergone maritime testing with many countries' armed forces in the last few years and is currently being used or tested by the armed forces of Austria, China, Egypt, France, Germany, India, Italy, Jordan, Libya, Pakistan, Spain, the United Arab Emirates, and the US.

Furthermore, as discussed above, some other actors, most importantly China, are pushing into the global UAV market.

Procuring off-the-shelf UAVs has many advantages. Most importantly, ready-made UAVs are immediately available and combat proven. However, not being involved in the development process can carry risks for the importer. In May 2013, the German Bundeswehr cancelled a Global Hawk deal (the EuroHawk variant) worth more than €0.5 billion citing significant barriers to certificating the UAV for flights in civil airspace (Gubisch, 2013). These difficulties were reportedly caused at least in part by Northrop Grumman's objection to sharing the details of the Global Hawk's construction plans with the German authorities (Die Zeit, 2013).

International joint ventures

States interested in acquiring an unmanned military capability can also opt for establishing joint ventures between domestic firms and foreign UAV manufacturers. Such collaborations are attractive as they ideally allow fast procurement of existing UAVs, which are modified in a way to fit the customer's needs. Depending on the type of joint venture they furthermore create employment and expertise within the country. Examples of such collaborations include the French Harfang, a surveillance UAV developed by IAI and EADS (*Defence Industry Daily*, 2013) and the British

Watchkeeper jointly developed by Elbit and Thales (*International Defense Review*, 2004). Joint ventures also exist between smaller UAV manufacturers such as Egypt's Ariesgsa Engineering Group and Argentina's Nostromo (Higuera, 2007), between the Vietnamese VASA and Russia's Irkut, VASA and the Unmanned Systems Group from Sweden (Grevatt, 2013) as well as between Italy's Selex Galileo and Pakistan Aeronautical Complex (airforce-technology.com).

However, joint ventures do not always lead to products combining the best of both worlds. The IAI/EADS Watchkeeper UAV was initially scheduled to enter service in 2006. At the end of 2013, the Watchkeeper is still 'inch[ing] towards release to service' (de Larrinaga, 2013c). Problems with software integration, flight trials, and certification have plagued the programme from the start. The US–German Euro Hawk project mentioned above, which strictly speaking is a cooperation between Northrop Grumman – which delivers the airframe – and EADS/Cassidian – in charge of the signals intelligence mission system – had eventually to be cancelled after expenses of more than €0.5 billion (Becker and Gebauer, 2013).

UAVs – a new military status symbol

How can the intensive proliferation of UAVs be explained? Why are so many countries now following Israel's and the US' lead and procuring drones?

States have emulated one another throughout history and have adopted other states' best practices, in military matters as well as in other areas. Emulation, it is believed, is easier than independent innovation and carries less risk as the new method or technology has already proven its worth elsewhere.

Neo-realists would most likely attribute the diffusion of UAV technology to drones' military worth and their versatility. UAVs, they would argue, have proven their worthiness in military operations and are therefore procured globally. UAVs improve reconnaissance and help to make ISR (intelligence, surveillance, and reconnaissance) missions easier and cheaper. The unmanned systems can be used for dull, dirty, and dangerous tasks. The employment of UAV models which can be controlled over long distances furthermore eliminates the risk to the pilot. Intelligence collected by UAVs helps to increase the precision of attacks, whether carried out by the UAV itself or by other means (UK MoD, 2011: Concl-3). The comparative cheapness of the technology and the ease with which it can be procured add to this. Accordingly, the global proliferation of UAVs should be nothing surprising.

This view, however, cannot fully explain the immense proliferation of UAVs in recent years. In particular, neo-realists such as Joao Resende-Santos (2007) argue that states, in order to improve their military effectiveness, will seek to copy the victor of the most recent (great) war and base the decision of what to emulate on its proven success. However, there is surprisingly little evidence for UAVs' military successes.

This does not mean that UAVs are not useful in military operations – indeed, there is also little evidence proving that UAVs are not effective. In fact, the (at times extremely heated and emotively conducted) debate on UAVs stands on surprisingly shaky ground. There is alarmingly little scholarship on the military use and worth of UAVs, in particular with regard to reconnaissance UAVs. The situation is equally obscure regarding armed drones; the efficiency of their use is strongly contested. While this in itself does not negate the possibility that UAVs are indeed effective and useful, it undermines the reasoning of the neo-realist approach. Also, few would consider the US – the most important UAV user to date – as clear winners of the wars in Iraq and Afghanistan.

Therefore, other factors appear to be driving the global diffusion of UAVs. One possible reason for the increased interest in the technology is rarely discussed: UAVs are on the way of becoming military 'must-haves', the symbols of a modern military.

'Drones are in vogue. No state that thinks highly of its military wants to do without the flying killers', writes the German *Spiegel* (Becker, 2013). Former German Defence Minister Thomas de Maizière voiced a similar opinion in a speech on UAVs in the Bundestag: 'We cannot say "we'll keep the stagecoach" while all others are developing the railway' (Bundestag, 2013: 27109). UAVs, according to this interpretation, are the future – those who are not part of it will lose out. French Air Force Brigadier General Patrick Charaix believes:

> If [France] wants to remain powerful within a coalition, we need to bring an unmanned capability to the table. Indeed, those countries that count have this military means which contributes on the one hand to the success of a mission and on the other hand increases the power and influence of the country.
>
> (Nijean, 2011)

Drones are 'hot'; they are a symbol of modernity, technological sophistication, and military superiority. A modern military, it seems, needs UAVs to be part of the 'big-boys club': 'Unmanned Aerial Vehicles have become a must-have item for modern armies' (*Defence Industry Daily*, 2012).

Accordingly, because of this belief, in many countries the acquisition of UAVs is handled at the highest political level. Hugo Chavez unveiled Venezuela's first UAV (BBC, 2012) and Iran's then-President Mahmoud Ahmadinejad proudly presented to the world the first Iranian 'unmanned bomber aircraft' (Express, 2010).[5] Russian President Vladimir Putin proclaimed that Russia would invest $12 billion on UAVs before the year 2020.

Political and military leaders around the world appear increasingly convinced that a self-respecting military needs unmanned (aerial) vehicles and therefore promote the acquisition of said technology. 'It's a prestige thing', argues Micah Zenko, analyst at the Council on Foreign Relations (Zenko, 2013a). This makes in particular the acquisition of smaller, cheaper, so-called 'tactical' UAVs attractive. They are comparatively cheap, easy to buy or build, and easy to handle and therefore do not require extensive (and expensive) training. At the same time, the procurement of a few of these small aircraft – which bear a striking resemblance to model toy airplanes – allows armed forces around the world to tick the box 'UAV holder' and to portray themselves as modern militaries. In fact, the acquisition of comparably small numbers of tactical UAVs by several countries points towards a 'box-ticking' exercise more than to a real military need. It may also be a sign that countries are cautious as UAVs' military utility is not yet non-ambiguously proven. Ultimately, even if adding a handful of tactical drones to a state's arsenal might not do much to increase its military capability, the government can still show off its new technological gadgets.

The effect of perceptions and beliefs with regard to military matters has been discussed by authors such as Theo Farrell, who proposes a cultural approach to the study of military innovation. Farrell studies the emulation of two 'transnational military norms' – conventional warfare and civilian supremacy – in the Irish armed forces (Farrell, 2001). He shows how both norms were emulated by the Irish military even though they were not necessarily beneficial to them.

The view prescribing the acquisition for UAVs is arguably not (yet) as strong as the transnational norms Farrell describes. Nevertheless the cultural approach offers an attractive alternative to the rationalist approach of neo-realists and deserves further attention.

UAV arms control

Journalists and anti-drone activists have expressed concern in view of the rapid global UAV proliferation and warn of an impending international arms race (Roberts, 2013). American political activist Barbara Ehrenreich claims that the 'insane proliferation' of UAVs poses 'nightmare possibilities' (Benjamin, 2013: viii). Others worry about the precedents established by the CIA's use of armed UAVs (Shane, 2011). Most criticism centres on the diffusion of armed UAVs rather than surveillance UAVs. Only few are alarmed by the – currently more significant numerically – proliferation of unarmed UAVs. Michael Boyle is concerned that the spread of drones will undermine 'many of the traditional concepts which have underwritten stability in the international system', such as deterrence (Boyle, 2013: 24). States may be less restrained to send unmanned aircrafts over their enemies' territories to spy on them, thereby creating the potential for international incidents (ibid.). On the other hand, an increased knowledge of opponents' military capabilities and movements through drone reconnaissance could also decrease the risk of dangerous misunderstandings and military conflicts caused by miscalculations (Franke, 2013).

Existing arrangements

For the time being, no international agreement or treaty exists which specifically targets UAVs, their transfer, sale, export, or use. UAV trade is however regulated by international legal regimes that predate UAV development, such as the Missile Technology Control Regime (MTCR).[6]

The MTCR is a regime established to regulate missile proliferation. It controls the transfer of equipment and technology which could contribute to the construction of missiles capable of delivering weapons of mass destruction. UAVs as defined by the MTCR include cruise missiles, target drones, reconnaissance drones, and other forms of armed or unarmed UAVs (National Security and Foreign Affairs Subcommittee, 2010). However, only UAVs meeting certain range/payload criteria are MTCR-controlled: the regime subjects exports of UAVs capable of delivering a payload of at least 500 kg to a range of at least 300 km and directly associated technology to an unconditional 'strong presumption of denial'.[7] Exports of complete guidance sets and warhead safing/arming/fusing/firing subsystems useable in such UAVs, and their directly associated technology, also are subject to a strong presumption of denial.

The MTCR has 34 states as members (including all G-7 States but not Israel) and is fundamentally dependent on their cooperation and vigilance in enforcing its guidelines. Therefore, it is constantly caught between states' desire to limit proliferation and the interest of domestic businesses seeking to increase exports. Accordingly, as the largest UAV producer, the US has in the past pushed five provisions that 'would have resulted in moving some UAVs currently categorized under MTCR Category I to Category II', making them easier to export (United States Government Accountability Office, 2012: 20). The proposals were rejected by the other MTCR members.

UAV export is furthermore in part regulated by the Wassenaar Arrangement which governs the export of conventional arms and associated dual-use items (Gormley and Speier, 2003: 73/74). Neither the MTCR nor the Wassenaar Arrangement directly address the proliferation of drone technology, but address them as a means of missile delivery or the dual-use technologies employed by drones.

Furthermore, most nations have domestic export controls regulating the sale of military items and civil/dual-use items. In the United States, these controls are administered by the State Department and the Department of Commerce (Gormley and Speier, 2003: 73). While the sale of UAVs to the UK and Italy have been authorised, Pakistan, the UAE, and Saudi Arabia have been refused drone technology by congressional restrictions (Boyle, 2013: 23).

UAV opposition and calls for international treaties and bans

A considerable number of anti-drone groups have been formed in recent years, most notably in the US and the UK.[8] Most groups focus their efforts on protesting against the United States' use of armed UAVs in targeted killings and for so-called 'signature strikes', in particular in the Pakistani tribal areas (Codepink, 2013), as well as against the prospects of autonomous UAVs which would be able to select and engage targets autonomously (Campaign to Stop Killer Robots, 2013).

Some of these organisations lobby for bans on certain types of UAV technology or for international treaties regulating their use (Rayman, 2013). For the moment, however, there is no agreement among politicians and experts on whether there is a need for such efforts and on what basis a ban should be established. While anti-drone groups express, for instance, moral concerns regarding the use of unmanned aerial vehicles in warfare, philosopher Bradley Strawser makes a convincing case regarding the moral duty military commanders have to protect their soldiers, thus arguing in favour of using UAVs (Strawser, 2010).

Most importantly, there are currently no incentives for states to lobby for a ban of UAV technology or for increased controls. UAVs appear to be useful military tools. Moreover, as shown in this chapter, UAVs have become a 'must-have' for modern militaries. A ban of UAV technology therefore seems unlikely and increased export controls are only desired to the extent that they do not hinder the ability of one's own national manufacturers to export UAVs. The only area in which international agreement appears currently possible is regarding the use of fully autonomous UAVs and robots. UN Special Rapporteur on extrajudicial, summary or arbitrary executions, Christof Heyns called for 'moratoria on lethal autonomous robotics' (UN Human Rights Council, 2013: 22); in November 2013 an agreement was reached to discuss the adding of a sixth protocol to the Convention on Conventional Weapons (CCW) which would result in a ban of these types of weapons.

Conclusion – UAVs in the future

This chapter has provided an overview of current UAV proliferation trends. The situation of the most important UAV nations has been analysed and current UAV arms controls presented. It has been suggested that a cultural analysis of UAV proliferation may be able to offer an explanation for the fast and extensive diffusion of UAV technology around the globe. With little scholarship to support the view that UAVs have proven their military worth and are therefore popular military goods, it has been argued that in addition to the neo-realist argument that weapons proliferate because they are useful, another factor seems to be at play. It has been shown that UAVs are becoming military 'must-haves', not first and foremost because of their usefulness but because the view is developing that militaries need UAVs in order to appear modern.

To be clear – this author does not negate the enormous military potential of UAVs. She also does not dismiss other factors which have been shown to influence military procurement, such as inter-service rivalries, bureaucratic quarrels, or regional rivalries. Nevertheless, as none of these factors can by itself explain UAV proliferation, this article offers an additional factor to be taken into account.

It is to be recommended that those interested in UAVs keep an eye on the cultural aspect of UAV procurement and use. Conceivably, norms will emerge regarding the use of UAVs. Many authors and anti-UAV activists have for instance expressed the fear that the (possibly illegal) US use of armed UAVs for targeted killings outside of official warzones could set a precedent for other countries to follow (Boyle, 2013; Roberts, 2013). But also other, less controversial, ways of employing UAVs, armed as well as unarmed, could lead to norms being established. Armed forces around the world are currently learning how to use UAVs in the most effective way. Doctrinal thinking and writing on the topic has begun (UK MoD, 2011; US Army, 2010). More research is therefore needed on current military UAV use in order to identify such norms early on. Further research is also needed with regard to the development of transnational norms in the military realm. What actors can play which roles? What prospects are there for actors outside the military to contribute to the development of these norms – particularly those at the receiving end of intervention?

Ultimately, international interest in UAVs is unlikely to abate and UAVs will spread further in the coming years. Even though opposition to the technology or certain aspects of its use is rising, for the time being a ban of the technology seems inconceivable, the establishment of meaningful international treaties improbable.

Notes

1 The term drone was originally used to describe pre-programmed UAVs whose flight path cannot be changed, while RPV describes UAVs piloted remotely from the ground or from another aircraft. These differences have, however, largely disappeared by now, mainly because most modern UAVs are both flown remotely and follow a pre-programmed flight path. NATO opted for a surprising half-measure by abandoning RPV but retaining the terms drone and UAS (NATO Standardization Agency, 2013).

2 Some elements of this section have been published previously (in German) in Franke, 2013 and Franke and Leveringhaus, 2014.

3 The nEUROn project is a pan-European endeavour involving France (Thales), Greece (Hellenic Aerospace Industry), Italy (Alenia Aermacchi), Spain (EADS CASA), Sweden (Saab), and Switzerland (RUAG). The aircraft is powered by a Rolls-Royce engine (de Larrinaga, 2012).

4 There is no commonly agreement on UAV classification system; sometimes the Hermes 450 is classified as tactical rather than MALE UAV, while some authors classify the Heron TP as HALE UAV.

5 It is doubtful whether this UAV has all the capabilities claimed.
6 For a more detailed discussion of control regimes applicable to UAVs, see Altmann, 2013.
7 www.mtcr.info.
8 US groups: The Upstate NY Coalition to Ground the Drones and End the Wars, http://upstatedroneaction.org; No Drones Network, http://nodronesnetwork.blogspot.co.uk; Code Pink, http://droneswatch.org/about. UK-based groups: Drone Wars UK, http://dronewars.net; Drone Campaign Network, http://dronecampaignnetwork.wordpress.com/about.

References

Adamsky, D. and Bjerga, K.I. (2010) 'Introduction to the Information-Technology Revolution in Military Affairs', *Journal of Strategic Studies*, 33, 4: 463–8.

AeroVironment (2013) 'UAS: RQ-11B Raven', www.avinc.com/uas/small_uas/raven (accessed 12 January 2014).

airforce-technology.com; Falco Tactical Unmanned Aerial Vehicle (UAV), Pakistan, www.airforce-technology.com/projects/falco-uav (accessed 12 January 2014).

Altmann, J. (2013) 'Arms Control for Armed Uninhabited Vehicles: An Ethical Issue', *Ethics and Information Technology*, 15, 2: 137–52.

Arthur, G. and Hardy, J. (2012) 'Pakistan Flexes UAV Muscles at Ideas', *Jane's Defence Weekly* (14 November 2012).

Axe, D. (2012) 'Russian Drones Lag U.S. Models by 20 Years', *Wired Dangerroom* (8 June 2012).

BBC (2012) 'Chavez Unveils Surveillance Drone' (14 June 2012), www.bbc.co.uk/news/world-latin-america-18446112 (accessed 14 January 2014).

Becker, M. (2013) 'Wettrüsten bei Unbemannten Flugobjekten: USA Verlieren Killerdrohnen-Monopol', *Spiegel Online* (27 March 2013), www.spiegel.de/wissenschaft/technik/uav-usa-verlieren-ihr-monopol-auf-bewaffnete-drohnen-a-885048.html (accessed 12 January 2014).

Becker, M. and Gebauer, M. (2013) '"Euro Hawk" wird zum 500-Millionen-Euro-Fiasko', *Spiegel Online* (14 May 2013), www.spiegel.de/wissenschaft/technik/euro-hawk-drohne-wird-zum-500-millionen-euro-fiasko-a-899812.html (accessed 14 January 2014).

Bell, M. (2012) 'Russia Makes Indigenous UAV Investment', *Jane's Defence Industry* (19 January 2012).

Benjamin, M. (2013) *Drone Warfare: Killing by Remote Control*, rev. edn, London/New York: Verso Books.

Binnie, J. (2012) 'Rise of the Pahphad: Iran's Unmanned Aerial Vehicles', *Jane's Defence Weekly* (21 November 2012).

Boyle, M. (2013) 'The Costs and Consequences of Drone Warfare', *International Affairs*, 89, 1: 1–29.

Bundestag (2013) 'Ausrüstung der Bundeswehr mit Bewaffneten Drohnen', Berlin (1 February 2013), http://dipbt.bundestag.de/dip21/btp/17/17219.pdf#P.27108 (accessed 14 January 2014).

Campaign to Stop Killer Robots (26 April 2013) 'Campaign Launch in London', www.stopkillerrobots.org/2013/04/campaign-launch-in-london (accessed 14 January 2014).

Codepink (2013) 'CODEPINK Peace Delegation to Pakistan', www.codepink4peace.org/article.php?id=6174 (accessed 13 January 2014).

de Larrinaga, N. (2012) 'Maiden Flight for Europe's Neuron UCAV', *Jane's Defence Weekly* (3 December 2012).

de Larrinaga, N. (2013a) 'BAE to Fly Taranis in 2013 as Potential Anglo-French UCAV Tie-up Emerges', *Jane's Defence Weekly* (21 June 2013).

de Larrinaga, N. (2013b) 'France Conducts its First MQ-9 Reaper Flight', *Jane's Defence Weekly* (26 September 2013).

de Larrinaga, N. (2013c) 'UK's Watchkeeper UAV Inches towards Release to Service', *Jane's Defence Weekly* (8 October 2013).

Defence Industry Daily (2012) 'Après Harfang: France's Next High-End UAVs' (21 February 2012).

Defence Industry Daily (2013) 'France's Harfang/SIDM IUAV Program' (18 February 2013), www.defenseindustrydaily.com/frances-harfang-sidm-iuav-program-06091 (accessed 12 January 2014).

Defense Science Board (2012) 'The Role of Autonomy in DoD Systems; Department of Defense' (July 2012), www.acq.osd.mil/dsb/reports/AutonomyReport.pdf (accessed on 12 January 2014).

Deutscher Bundestag (2013) 'Plenarprotokoll 17/219', Stenografischer Bericht, 219, Sitzung, Berlin.

Die Zeit (2013) 'Drohnenhersteller Widerspricht de Maizière', *Die Zeit* (22 May 2013), www.zeit.de/politik/deutschland/2013–05/drohne-euro-hawk-hersteller (accessed 14 January 2014).

Engelhardt, T. (2009) 'Drone Race to a Known Future, Why Military Dreams Fail – and Why It Doesn't Matter', *TomDispatch* (10 November 2009), www.tomdispatch.com/post/175155 (accessed 12 January 2014).

Express (2010) 'Iran President Unveils Drone Bomber' (23 August 2010), www.express.co.uk/news/world/194950/Iran-president-unveils-drone-bomber (accessed 12 January 2014).

Farrell, T. (2001) 'Transnational Norms and Military Development: Constructing Ireland's Professional Army', *European Journal of International Relations*, 7, 1: 63–102.

Franke, U.E. (2013) 'Verbreitung von Unbemannten Flugzeugen für den Militärischen Gebrauch', *Aus Politik und Zeitgeschichte*, APuZ 37, 2013: 33–8.

Franke, U.E. and Leveringhaus, A. (2014) 'Militärische Robotik', in Jaeger, T. (ed.) *Handbuch Sicherheitsgefahren*, Wiesbaden: Verlag Springer (forthcoming).

Frost and Sullivan (2013) 'Israel is Top Global Exporter of Unmanned Aerial Systems with a Continued Positive Outlook Ahead', *Frost & Sullivan* (20 May 2013), www.frost.com/prod/servlet/press-release.pag?docid=278664709 (accessed 12 January 2014).

Garamone, J. (2002) 'From the U.S. Civil War to Afghanistan: A Short History of UAVs', *American Forces Press Service* (16 April 2002).

Gormley, D.M. and Speier, R. (2003) 'Controlling Unmanned Air Vehicles: New Challenges', *The Nonproliferation Review*, 10, 2: 66–79.

Grevatt, J. (2013) 'Vietnam's Unmanned Ambitions', *Jane's Defence Weekly* (6 August 2013).

Gubisch, M. (2013) 'Germany Pulls Plug on Euro Hawk UAV Programme', *Flight Global* (16 May 2013).

Harshberger, T. (2012) 'Expect More Drone Use Like Recent Israeli Episode', *Rand Corporation*, www.rand.org/commentary/2012/10/11/USNEWS.html (accessed 12 January 2014).

Hastings, M. (2012) 'The Rise of the Killer Drones: How America Goes to War in Secret', *Rolling Stone* (16 April 2012).

Hewson, R. (2013) 'Variation on a Theme: China Gets Real on UAV Exports', *International Defence Review* (15 March 2013).

Higuera, J. (2007) 'Argentina, Egypt Collaborate on Unmanned Systems', *Jane's Defence Weekly* (28 June 2007).

Holland, J. (2012) 'NATO awards $1.7 bn Global Hawk AGS contract to Northrop Grumman', *Jane's Defence Weekly* (21 May 2012).

Hsu, K. (2013) 'China's Military Unmanned Aerial Vehicle Industry', *US-China Economic and Security Review Commission. Staff Research Backgrounder* (June 2013).

International Defense Review (2004) 'Thales Chosen for Watchkeeper', *International Defense Review* (17 September 2004).

IISS (International Institute for Strategic Studies) (2000) *The Military Balance*, London: IISS.

IISS (International Institute for Strategic Studies) (2008) *The Military Balance*, London: IISS.

Jane's Defence Weekly (2007) 'Iran Claims Production and Use of "Stealth" UAV', *Jane's Defence Weekly* (16 February 2007).

Jennings, G. (2010) 'Predator-Series UAVs Surpass One Million Flight Hours', *Jane's Defence Weekly* (8 April 2010).

Lorenz, A., von Mittelstaedt, J. and Schmitz, G.P. (2011) 'Waffen: Botschafter des Todes', *Spiegel* Online 2011/42, www.spiegel.de/spiegel/a-791992.html (accessed 12 January 2014).

Macleod, H. and Flamand, A. (2012) 'Are Iran's Drones Coordinating Attacks in Syria?', *Global Post* (18 September 2012), www.globalpost.com/dispatch/news/regions/middle-east/syria/120914/iran-drones-syria (accessed 12 January 2014).

Mazzetti, M. (2013) *The Way of the Knife*, New York: Penguin.

McDaid, H. and Strong, B. (2003) 'Remote Piloted Aerial Vehicles: An Anthology', www.ctie.monash.edu.au/hargrave/rpav_home.html#Beginnings (accessed 12 January 2014).

Michaels, J. (2013) 'Experts: Drone Basis for New Global Arms Race', *USA Today* (9 January 2013).

Ministry of Defence (2011) 'The UK Approach to Unmanned Aircraft Systems', *Joint Doctrine Note* 2/11 (30 March 2011).

Minnick, W. (2012) 'China's Unmanned Aircraft Evolve from Figment to Reality', *Defense News* (26 November 2012), www.defensenews.com/article/20121126/DEFREG03/311260001 (accessed 12 January 2014).

Mtcr.info 'Guidelines for Sensitive Missile-Relevant Transfers', Missile Technology Control Regime, www.mtcr.info/english/guidetext.htm (accessed 12 January 2014).

Munson, K. (2007) 'UK RAF Expects First Predator B Deliveries', *Jane's Defence Weekly* (27 March 2007).

National Security and Foreign Affairs Subcommittee of the Committee on Oversight and Reform (2010) 'Testimony by Kevin J. Wolf Assistant Secretary for Export Administration Bureau of Industry and Security U.S. Department of Commerce' (23 March 2010), www.google.co.uk/url?sa=t&rct=j&q=&esrc=s&source=web&cd=1&ved=0CDAQFjAA&url=http%3A%2F%2Fwww.bis.doc.gov%2Findex.php%2Fforms-documents%2Fdoc_download%2F730-testimony-by-assistant-secretary-wolf-before-the-national-security-and-foreign-affairs-subcommittee&ei=FmfSUtSDBsjD7Aai8oDoBg&usg=AFQjCNHw_eMwMopXSJ32oAEyXDzG6SxpFw

&sig2=IbDtgTvpKaHg5b_g4U8pbA&bvm=bv.59026428,d.d2k (accessed 12 January 2014).

NATO Standardization Agency (2008) 'NATO Glossary of Terms and Definitions', English and French, AAP-6, www.fas.org/irp/doddir/other/nato2008.pdf (accessed 12 January 2014).

Nijean, J.L. (2011) 'Colloque International: En Route Vers un Drone Armé (8 June 2011), www.defense.gouv.fr/air/actus-air/colloque-international-en-route-vers-un-drone-arme (accessed 12 January 2014).

Panetta, L. (2009) 'Obama's Middle East Strategy, AFPAK Drone Strikes Are Only Game in Town', *New Perspectives Quarterly* (1 June 2009).

Paraszczuk, J. (2012) 'Iran Admits Exporting Drone Tech to Venezuela', *Jerusalem Post* (12 December 2012).

Rayman, N. (2013) 'The Campaign to Kill "Killer Robots" Gains Steam', *Time* (25 October 2013), http://world.time.com/2013/10/25/the-campaign-to-kill-killer-robots-gains-steam (accessed 12 January 2014).

Resende-Santos, J. (2007) *Neorealism, States, and the Modern Mass Army*, Cambridge: Cambridge University Press.

Roberts, K. (2013) 'When the Whole World Has Drones', *National Journal* (22 March 2013).

Shane, S. (2011) 'Coming Soon: The Drone Arms Race', *New York Times* (8 October 2011).

Sloggett, D. (2012) 'Iranian Reconnaissance of Dimona?', *Air Forces Monthly* (December 2012), 20–2.

Spinetta, L. (2010) 'The Rise of Unmanned Aircraft', *Aviation History Magazine* (10 November 2010).

Strawser, B.J. (2010) 'Moral Predators: The Duty to Employ Uninhabited Aerial Vehicles', *Journal of Military Ethics*, 9, 4: 342–68.

Taylor, A. (2012) 'Russia Is Planning to Spend $12 Billion on Drones before 2020', *Business Insider* (14 June 2012), www.businessinsider.com/russia-putin-drones-2012-6#ixzz2IvfDXpPH (accessed 12 January 2014).

Teal Group Corporation (2013) 'Teal Group Predicts Worldwide UAV Market Will Total $89 Billion in Its 2013 UAV Market Profile and Forecast' (17 June 2013), http://tealgroup.com/index.php/about-teal-group-corporation/press-releases/94-2013-uav-press-release (accessed 12 January 2014).

Trautermann, M. (2006) 'Unmanned Aerial Vehicles, Katalysator für die Transformation', *Europaische Sicherheit*, 55: 49–56.

UK Ministry of Defence (2011) 'The UK Approach to Unmanned Aircraft Systems', *Joint Doctrine Note* 2/11 (30 March 2011).

UN Human Rights Council (2013) 'Report of the Special Rapporteur on Extrajudicial, Summary or Arbitrary Executions', Christof Heyns, United Nations A/HRC/23/47 (9 April 2013).

US Army (2010) 'UAV Roadmap 2010–2035' (April 2010), www-rucker.army.mil/usaace/uas/US%20Army%20UAS%20RoadMap%202010%202035.pdf (accessed 14 January 2014).

US Government Accountability Office (2012) 'Nonproliferation. Agencies Could Improve Information Sharing and End-Use Monitoring on Unmanned Aerial Vehicle Exports', GAO-12-536 (July 2012).

Wan, W. and Finn, P. (2011) 'Global Race on to Match U.S. Drone Capabilities', *Washington Post* (4 July 2011).

Williams, H. (2011) 'Camcopter Undertakes Shipborne Trials with French Navy', *International Defence Review* (5 December 2011).

Wilson, J.R. (2011) *Worldwide UAV Roundup*, Aerospace America (March 2011). Complete chart available at www.aerospaceamerica.org/Documents/March%20 2011%20AA%20PDFs/UAV_CHART_2011.pdf (accessed 14 January 2014).

Wolfgang, B. (2013) 'Drone Industry Gives Journalists Not-so-subtle Hint: Don't Use the Word "Drones"', *Washington Times* (14 August 2013), www.washington-times.com/news/2013/aug/14/drone-industry-journalists-dont-use-word-drones/?utm_source=dlvr.it&utm_medium=appnet (accessed 14 January 2014).

4 The political appropriation of casualties in threat construction

The case of US drone strikes in Pakistan

Wali Aslam[1]

Introduction

This chapter examines how the casualties of American drone strikes in Pakistan have been politically appropriated by two particular actors in order to construct the perception of threat socially. On one side, a discourse analysis of the speeches and statements of Imran Khan, leader of the Pakistan Movement for Justice opposing these strikes, is conducted. On the other, United States political and military-intelligence officials' statements supporting these strikes are studied to show how casualties are deployed politically to support opposing arguments. The chapter argues that threat construction is a social process that involves opposing sides appropriating the circumstances around drone casualties and deploying their deaths and injuries to represent a threat and shape a reality through their construction of a particular discourse. That process involves active participation by political elites and a receptive audience. Once constructed this way, the nature of these constructed threats then proceeds to play an integral role in shaping the political strategies of each side.

The American drone attacks in the tribal areas of Pakistan have been covered by authors from different perspectives, including critical (see Aslam, 2011) and ethical (see Enemark, 2011) evaluations of the practice as well as the study of its efficacy (Taj, 2010).[2] However, the literature on the subject does not ask the wider question: how do different actors involved in the process appropriate the casualties of the drone attacks to construct and reinforce their version of a threat? The question is pertinent because actors supporting or opposing the drone strikes justify their actions by referring to those killed or injured as a result. Those opposing the practice point out that the dead are innocent victims of attacks and that the United States should be held responsible for these killings (Woods, 2012b). On the other hand, actors supporting the attacks also refer to the dead in their political discourse, but do so by pointing out that the drones usually kill only terrorists and that their collateral damage is minimal (Ross, 2013).

Interestingly, both sides refer to the same phenomenon, the killings of individuals by unmanned drone aircraft, but each appropriates the phenomenon to justify its own actions and to construct a perception of an external threat. In other words, the very same phenomenon is 'hijacked' by two opposing sides for very different aims, and through the process of securitisation a threat is constructed that favours the agenda of those constructing the threat. This practice then goes on to justify not only the future policy decisions in the United States but also the reactions to drones by political actors in Pakistan. Hence this chapter studies the phenomenon of political appropriation of the drone casualties by actors on different sides of the divide in order to analyse the process through which a threat is socially constructed. It employs the theoretical approach of Social Constructivism to conduct its task.

The chapter begins by presenting and discussing the theoretical principles relating to the social construction of a 'threat' that are deduced from Social Constructivism. Afterwards, it introduces Imran Khan, a major political actor in Pakistan politics. Khan retired as a world-class cricketer to lead a political party in Pakistan, the *Pakistan Tehrik-e-Insaf* (PTI, or Pakistan Movement for Justice). The study uses Imran Khan since his party's popularity has increased substantially in recent years. According to a recent survey conducted by the International Republican Institute (IRI), a Washington based think-tank, it is now the second-most popular political party in the country (Masood, 2013). The PTI was successful in forming a government in the Khaibar Pakhtunkhawa province of Pakistan (KPK) as a result of the country's recent general elections. The party also secured the second-highest number of votes in Pakistan on a national level and is now the third largest party in the country's National Assembly (Dawn, 2013). The party has vowed to act as a vocal opposition to the ruling Pakistan Muslim League (N) (PML (N)) party of Nawaz Sharif and its opinion will be influential inside as well as outside of the country. After introducing Imran Khan, and his political party, the chapter proceeds to examine how Khan has been appropriating the drone casualties in order to construct what it conceives to be a threat: the presence of the United States inside Pakistan.

The discussion then moves on to introduce different supporters of the drones in the United States, focusing on the speeches given and documents published by various officials in the current Obama administration. It highlights how officials are appropriating the victims of drones to justify their stance supporting the global campaign against terrorism. The chapter also highlights the phenomenon of political appropriation by officials in the United States and how it eventually leads to threat construction.

The chapter's main premise is Constructivist in nature and it does not aim to contribute to any positivist debates on the practice of drone strikes. In other words, it does not aim to prove or disprove anything as such; it focuses on how social reality, and our perception of it, is constructed to play a crucial role in justifying the policies and actions of those opposing or supporting the use of drones. The chapter mostly focuses on the first term of President Obama and does not spend much time studying events from the George W. Bush administration. Despite the first-ever drone strike in Pakistan's Federally Administered Tribal Areas (FATA) taking place in 2004, for a cohesive analysis of the discourse of various actors involved it is useful to start with a limited time period. Additionally, this chapter dedicates itself to studying the drone phenomenon only in the tribal areas of Pakistan and does not discuss the strikes in other parts of the world such as Somalia and Yemen.[3]

Social construction of threats

The starting premise of Social Constructivism is that the world surrounding us is not an objective phenomenon; rather, it is constantly 'made and remade as human actors bring meanings to their interactions with each other' (Fierke, 2007: 56). This means that to construct something would be an *act* that brings a subject or an object into being that otherwise would not exist (ibid.; emphasis added). Like many other aspects of the reality around us, threats are socially constructed by human actors through their interaction in a social world and a specific context (ibid.). As such, saying that threats are the product of the process of social construction means that they would be 'imbued with social and cultural values, norms and assumptions rather than being the product of purely individual thought or meaning' (ibid.).

The process of creating threats through social construction has been termed 'Securitisation' by the Copenhagen School of Constructivism. Securitisation takes place when 'an issue is presented as posing an existential threat to a designated referent object' (Buzan *et al.*, 1998: 21). The public is told that the 'special nature of the security threats justifies the use of extraordinary measures to handle them' (ibid.). A study of military security looks at referent objects such as defence of the state and its armed forces. On the other hand, an examination of the political aspect of security focuses on the state's sovereignty and its ability to govern itself, either or both of which may be under threat from a perceived enemy (ibid.: 22). On a societal level, the referent objects of this process of secur_ itisation can be 'large scale collective identities' such as national or religious groups (ibid.: 22–3).

The process of securitisation takes place when a securitising actor 'articulates an already politicised issue as an existential threat to a referent object' (Emmers, 2007: 111). Given the existential nature of the threat, the securitising actor favours adopting extraordinary means beyond the day-to-day responses to threats (ibid.: 111). Securitisation cannot be understood, however, without focusing on the issue of identity. Certain objects are designated as threats or enemies not on the basis of their objective nature as threats, but only with reference to the identity of those threatened. Buzan *et al.* (1998: 23) argue that quite often challenges and changes are portrayed as 'threats to identity, because "we will no longer be us", no longer the way we were or the way we ought to be to be true to our "identity"'. Identities are important in providing stability in our lives as they ensure predictable patterns of behaviour among actors (Hopf, 1998: 174). 'Identities perform three necessary functions in a society: they tell you and others who you are and they tell you who others are' (ibid.: 175). In telling an individual who they are, identities 'imply a particular set of interests or preferences with respect to choices of action in particular domains, and with respect to particular actors' (ibid.: 175).

The notion of identity is dependent on 'historical, cultural, political, and social context' (ibid.: 176). By extension, the interests of states, or actors within these states, are products of their identity. The 'identity "great power" implies a particular set of interests different from those implied by the identity "European Union member"' (ibid.: 176). To examine issues of identity requires asking which group needs to be protected, by whom and from what threats (MacDonald, 2005: 299). It is also closely linked to highlighting the group's values: what it holds dear and what it would like to put forward as its most precious possession. Similarly, answering the question of who will do the job of protecting against a particular threat would by default grant legitimacy to those driving the process of securitisation. An analysis that looks at the most important threat in a given situation would lead to focusing on certain constructed threats within a given time and a social context while ignoring other issues that could be construed as threats.

Unlike Neorealism, Constructivist approaches dedicate considerable energy to studying the non-material factors involved in threat construction. In this regard, what is defined as an interest is obviously important. However, a better understanding of the process of threat construction can be gained by observing what is missing from the repository of interests and to ask why it is missing. Constructivists regard the missing interests as 'produced absences, omissions that are the understandable product of social practices' and 'the social practices that constitute an identity cannot imply interests that are not consistent with the practices' (Hopf, 1998: 176).

Norms also play a central role in the social construction of threats. Norms, defined as shared expectations of appropriate behaviour held by a community of actors (Finnemore, 1996: 22), are instrumental in classifying someone as a friend or an enemy. One's friends are supposed to act in a way that is the norm, or expected by everyone in a group to be an example of reasonable behaviour, while one's enemies act contrary to the way they may be expected to behave in international society.

Collins argues that the process of securitisation reveals the power and influence of the securitising actor (Collins, 2005; see also Emmers, 2007: 112). In this context, power relates to the ability of an actor to construct what constitutes 'common sense' (MacDonald, 2005: 300). In the process of threat construction, an actor will be described as powerful if they are able to define what constitutes 'good behaviour in international society' (MacDonald, 2005: 300). An actor's ability to securitise certain aspects of life is also linked to the possibility that they could increase their power through the use of the mechanisms at their disposal, thereby bringing into focus the 'political motives' of securitisation (Emmers, 2007: 117). The securitising actor does not always have to belong to a state; they can be political elites in a society independent of a state.

MacDonald and Merefield (2010: 188) argue that security gains meaning through the processes of 'contestations' and 'negotiations' in which 'communication is central; and it is intimately related to particular narratives of who those communities are and what they value'. The process of negotiation involves 'highly strategic actors recognising the need for public support and the need to undermine perceptions of the legitimacy of alternatives' (ibid.: 188). It is a process of articulation between the articulators and the audiences at whom the articulation is directed (MacDonald, 2008: 572). The idea of contestation, on the other hand, looks at how some narratives of identity and security are empowered and others are marginalised with multiple actors putting forward multiple visions (MacDonald and Merefield, 2010: 202). The issue of contestation is related to studying 'how different narratives of security and identity compete to provide the framework through which the world is understood and approached, and "our values" defended and advanced' (ibid.: 190).

The ideas of 'negotiation' and 'contestation' put discourse at the heart of threat construction. Wæver (1995: 55) regards security as a *speech act* – 'the utterance *itself* is the act' and 'by saying it something is done'. Buzan *et al.* (1998: 24) believe that the process of securitisation constructs a security threat not because of an existential threat's mere existence, but rather due to the way in which it is presented:

> A discourse that takes the form of presenting something as an existential threat to a referent object does not by itself create securitization – this is a securitizing move, but the issue is securitized only if and when the audience accepts it as such.
>
> (Ibid.: 25)

Furthermore, 'successful securitization is not decided by the securitizer but by the audience of the security speech-act' (ibid.: 31). Hence a key question is 'does the audience accept that something is an existential threat to a shared value?' (ibid.: 31). This highlights the intersubjective nature of the process of threat construction. Realists regard security as a platform for the state elites. Political elites construct threats and pass them on to the public without much participation from the public itself. Constructivism, on the other hand, asserts that securitisation is a two way process involving the audiences that legitimise the views of the elites who construct the threats.

The social construction of a threat very much depends on a 'shared understanding' of what constitutes a danger to security, and the process only requires the acknowledgement by audiences that a threat exists – and not the adoption of extraordinary means to tackle that threat (Emmers, 2007: 113–14; Buzan *et al.*, 1998: 25). Securitisation is context-dependent. It also depends on the psycho-social disposition of the audiences involved as well as the power that both a speaker and a listener bring to the interaction (Balzacq, 2005: 172). The context is important because it determines how the referent objects of security (individuals, groups, values held by a group) come to be threatened (MacDonald, 2002: 281).

Dotty argues that the type of discourse used by agents creates 'interpretive possibilities' that make it 'virtually impossible to think outside of it' (Dotty, 1993: 302). An approach that focuses on 'discursive practices as a unit of analysis can get at how this "reality" is produced and maintained and how it makes various practices possible' (ibid.: 303). In other words, discourse is very important to the social construction of threats. For speeches and press conferences to be taken seriously, they must make sense to the general public and must fit with what the public describes as 'reality' (ibid.: 303).

Pakistan Movement for Justice and Imran Khan

Imran Khan is better known across the world for his sporting successes than his political prowess. In Pakistan, he is famous as the man who led his country's cricket team to win the 1992 World Cup. Khan started his political party, the PTI, in 1996 (see Burke, 2012a). The PTI is a right-of-centre party that targets voters who would usually vote for PML (N).[4] Khan's political views on various internal and external policy matters have tended to be quite right-wing. For example, Khan has criticised the US for the violence Pakistan has witnessed in recent years, arguing that the US drone strikes have caused terrorism in the country; he has claimed that the only way militants react to the attacks by the US is by blowing up the innocent people of Pakistan (UK Channel 4, 2012). Khan's opponents have claimed that his party has received support from Pakistan's powerful military-intelligence establishment, which generally sits outside of the democratic sphere (*The Express Tribune*, 2011). The PTI had no major political following until it re-launched itself in October 2011 at a rally in Lahore that attracted around 100,000 people (Nelson, 2011). The success of this rally led to the PTI gaining a massive increase in the number of its followers. The rally sparked a move of many prominent leaders from different political parties in Pakistan, including the Pakistan People's Party and the PML (N), towards the PTI.

Political appropriation of casualties by Imran Khan

Khan has repeatedly referred to the US drones in various political rallies and media interviews (*The News*, 2013a). His rhetoric has focused on the issue of the Pakistani people's identity and their common cultural values, and his words are very much soaked in the traditional terminology of South Asian culture, incorporating notions such as honour and shame (*The Express Tribune*, 2012). A consistent theme in Khan's political rhetoric is that current Pakistani civilian leaders have brought shame to the country's name by not taking a stance against the use of drones by the United States (Nelson, 2011). He has referred to recent former Presidents of Pakistan Pervez Musharraf[5] and Asif Zardari[6] as 'American stooges' (Crilly, 2011). This sharp focus on the identity of the nation portrays the US as a threat as it is cast as an entity that is attempting to alter the idea of who Pakistanis are. Instead of comprising an honourable nation able to defend itself, Pakistanis are portrayed as helpless victims who are not able to protect their honour and who have no choice but to let a foreign enemy conduct attacks within their territory.

Khan has very deftly used the drones with reference to Pakistan's historical subservience to foreigners: first the British (as it was a British colony before 1947) and then the Americans. Playing on the public's desires to have an identity of an independent nation free in decision making (thus unaffected by foreign influence), he has used drones as a symbol of not only foreign attempts to subjugate Pakistan but also the inability of the 'puppet' rulers of Pakistan to stand up to the United States and resist such attacks (*The Nation*, 2013). A disdain for colonialism is a common theme in Khan's ideas and has been repeatedly registered in his book and elsewhere (see Khan, 2012b: 22 and 24). For Khan, drones are a symbol of colonisers and Pakistani rulers have 'sold national sovereignty for dollars' (*Daily Times*, 2011). Pakistani sovereignty and nationalism are a constant refrain in Khan's speeches and drones provide a useful opportunity to play on these themes of identity in order to securitise the current situation for Khan's political gain, as highlighted above. In 2011 Khan was quoted as saying: 'we believe drone strikes are illegal according to international law because they kill innocent people ... the U.S. or any other country has no right to violate frontiers of an independent state' (Yusufzai, 2012). The play on nationalism has resonated quite well with the youth of Pakistan, many of whom believe that the country could do much better if left to fend for itself (*Financial Times*, 2013). Use of this powerful and culture-heavy language is useful for Khan in portraying the United States as a threat. Indeed his views indicate that a continued acceptance of the strikes will make Pakistan a subservient nation which, as Buzan *et al.* (1998: 23) put it, will no longer be what it means to be Pakistani, no longer the way Pakistanis were or the way they ought to be to be true to their identity.

Interestingly, for Khan, the constant support of Pakistani intelligence agencies and the military establishment for disparate terrorists – action that in themselves may attract drone attacks – do not feature in his rhetoric (see Cockburn, 2010). They are the 'produced absences' discussed by Hopf (1998: 176). Of course, doing so may draw unwanted attention to the role of the Pakistani military and intelligence agencies in supporting the militant movement. They may be blamed for causing the deaths of innocent people as their policies are inviting the US to conduct the drone-related operations. Khan, as securitising actor, has omitted these material facts from his discourse as they do not fit the process of constructing the United States as a threat. Where the US is portrayed as violating Pakistani sovereignty, the audience at Khan's rallies is not informed of the violations of the country's sovereignty committed by Al-Qaeda and other militants hiding in the tribal areas, despite the fact that these individuals lack both visas and the permission of Pakistani authorities. Khan overlooks this information because, of course, revealing it would contradict the argument that the US is the actual threat to Pakistan – and not its own military-intelligence establishment nor the militants hiding in the tribal areas. Furthermore, Khan will not highlight the neglect of the Pakistani authorities in letting these militants hide in its territory, as doing so does not contribute to the narrative he wants to promote. In brief, for Khan, his ideology constructs a tunnel vision for him that focuses only on one aspect of the problem. It does not provide him space to look at how the Pakistani authorities themselves might be contributing to the problem that necessitates the use of drones by the United States.

Pakistan's former President, General Pervez Musharraf, elected to side with the US in its global war against terrorism following the 2001 attacks on the World Trade Centre in New York City. As part of this cooperation, the Pakistan army initiated a string of counterterrorism operations in the country's tribal areas, financed by the United States, which led to a substantial increase in violence in that part of the world. Khan argues that

> until that point [when Pakistan joined the US] we had no militant Taleban in Pakistan. We had militant groups, but our own military establishment was able to control them. We had madrassas, but none of them produced militants intent on jihad until we became a frontline state in the War on Terror.
>
> (Taqi, 2010)

For Khan, the tribal areas became a hotbed of anti-US and anti-Pakistan extremism as a result of Pakistan's decision to fight 'someone else's war' (*The News*, 2013b). The way in which this narrative has been constructed attributes the problems in the tribal areas to only one cause: the US involvement in the region. Seen from this perspective, it is not possible to think outside of the 'interpretive possibilities' discussed earlier, as for a common man or woman in Pakistan it is very easy to accept that there may be no other reason behind the tribal areas becoming a hub of militancy after the 9/11 attacks (Dotty, 1993: 302). Furthermore, playing on the familiar nationalist rhetoric leads to the discussion of certain policy choices Pakistan should make, which relate to distancing itself from America's war as it is the alliance with the US that has led the country to where it currently stands. This oversimplification of the actual situation suits Khan's agenda as it portrays an 'us versus them' state of affairs that is simple for the intended audience to grasp. It does not suit Imran Khan to discuss the nuances of whom the army is targeting and whom it is refusing to target, as doing so would focus the attention on the army's strategies and its role in attracting the US drones. It is now quite well-known that the Pakistan army has refused to conduct any counterterrorism operations in the North Waziristan agency where militants attacking the NATO and Afghan forces are hiding (Abouzeid, 2010).[7] Clearly, for Khan, scrutiny of Pakistan's policies in this regard is a venture that does not pay as many political dividends as a straight opposition to US operations in the region.

US officials have publicly stated that by targeting militants in the tribal areas, they are assisting Pakistan through eliminating elements that are a threat to Pakistan (Khan, 2008). However, Imran Khan sees no militant threat in the tribal areas that is as threatening for Pakistan as it might be portrayed by the US policymakers. Khan has on occasions argued that 'there is not a threat to Pakistan from Taliban ideology' (Burke, 2012b). This has for many liberal commentators cemented an image of Khan and the current leadership of PTI as a deeply conservative political extension of the ISI (Inter-Services Intelligence). The party has proven very effective at mobilising the Pakistani public by using such rhetoric that leads to an increase in the military's influence, giving credence to the idea that securitisation is always conducted to suit someone's agenda. In this case, the securitising actor (Imran Khan) is highlighting the Pakistan army's agenda through his rhetoric. The party provides a very useful political platform to the army that is being increasingly restricted from interfering in politics after the transition to democracy in 2008 (see Bukhari, 2012). Such rhetoric helps increase the military's bargaining power vis-à-vis the US; it can use the public rallies to give the country the message that it is doing its interests a big favour by supporting American policies despite strong resentment in Pakistan.

The decision to adopt a hard-line stance on drones is an influential unifying factor in a country that has become increasingly divided on sectional, sectarian, and ethnic lines. Instead of thinking of themselves as Sindhis, Punjabis, and Pathans, the audience addressed by Khan see themselves as Pakistanis – a group which, collectively, is the target of the US onslaught. Khan's rallies have primarily attracted crowds from Pakistan's energetic and politically disillusioned youth whose emotions are aroused by the idea that their country is being bombed through no fault of its own. Playing on (and perhaps playing up) those emotions, Khan has declared 'they're talking about a country in which 35,000 people have died during a war that had nothing to do with us. Ours is perhaps the only country in history that keeps getting bombed, through drone attacks, by our ally' (Jeffries, 2011). This politically fruitful toying with the emotions of the nation's youth has contributed to a recent strengthening of anti-US sentiment in Pakistan, causing further difficulties for the US in gaining the sympathies of the Pakistani people. Such rhetoric has also led to an increase in Khan's own popularity. Khan has been portraying himself as a leader who would be able to do what others are not: i.e. stand up to the US and stop its attacks on Pakistani soil. This is a classic example of an actor constructing a threat and then proceeding to say that nobody else but he can solve the problem, thereby conferring legitimacy on himself as a leader. Though it may be too early to say so definitively, this could very likely be a reason why Khan's party has swept into power in the KPK province – it borders the FATA territory, which has been the primary target of American drones.

Imran Khan has also been portraying the use of drones as a senseless American policy similar to the ones adopted by the US in the past. As discussed in section two, this is a way to portray the US as an irresponsible power that acts contrary to the norms of international society. For example, Khan has said:

> the war on terror is the most insane and immoral war of all time. The Americans are doing what they did in Vietnam, bombing villages. But how can a civilised nation do this? How can you eliminate suspects, their wives, their children, their families, their neighbours? How can you justify this?
>
> (Jefferies, 2011)

Such comparisons with Vietnam guide the audience's reactions to drones as a policy that goes against the idea of a 'good behaviour' in international society. The message is clear: if Washington continues to follow this approach, it will end up being a disaster like the Vietnam conflict. This simplistic comparison with Vietnam suits Khan even if the circumstances of two theatres may be substantially different.

In Khan's opinion, the use of drones must be stopped immediately as it can only lead to a worsening of the current situation. According to him, there are a million armed men in FATA (Khan, 2011) and they may be dragged into the 'senseless' war with the US if nothing is done in time to stop the country's actions. Pakistani officials have repeatedly stated that drones have been counterproductive in their policy to isolate Al-Qaeda-linked militants (*The Times of India*, 2011). Not surprisingly (and for reasons discussed above), Khan's rhetoric supports the views of the country's military establishment. Furthermore, he is able to instil those ideas in the masses, who have come to regard him as a leader quite different from the traditional leadership the country has had for a time.

Political appropriation of casualties in the United States

The previous section looked at how Imran Khan has managed to use the ongoing drone attacks in Pakistan to portray the US as a threat for its own political gain. This section will examine how officials in the US have employed the use of drones to justify a 'war paradigm' that necessitates certain US actions. The George W. Bush Administration defined the attacks of 11 September 2001 as the events that changed everything (Cheney, 2003; see also Hafetz, 2012: 844). In response to the attacks on the twin towers, the Bush Administration initiated a war on terrorism that was ostensibly global in reach. It would not be 'one battle but a lengthy campaign unlike any other we have seen' (Bush, 2001). The launching of the war on terrorism and using the 'war paradigm' instead of a simple law-enforcement framework were not mere rhetorical tools but rather have had significant implications on various fronts, such as the targeting of enemy combatants and the rights of detainees (Blank, 2011). These actions have 'facilitated and encouraged the growth of authority without the corresponding spread of obligation in many cases' (ibid.: 723).

The Obama Administration came into office in 2009 on the slogan of change. It subsequently dropped the use of the phrase 'war on terrorism' and replaced it with the slightly more cumbersome 'overseas contingency operations' (Burkeman, 2009). The new Administration's motivation for coining a different terminology was to distinguish its counterterrorism efforts from its predecessors with the intention of making a break from the past (Becker and Shane, 2012). However, a close scrutiny of the statements and speeches of the current Washington officials with reference to the use of American drones to target terrorists in Pakistan's tribal areas reveals that Washington has not only continued to justify its activities in terms of a 'war' but also provide more evidence of a continuity of the Bush-era policy than a real change from the past that was promised to the American people (see Ralph, 2010).

In his justification of the drones, President Obama stated that drones were 'part and parcel of our overall authority when it comes to battling Al-Qaeda' (quoted in Woods, 2012a). The use of battlefield analogy is closely linked to perpetuating the impression that America is *at war* with Al-Qaeda – this would, of course, necessitate a different response from mere law enforcement. Following in Obama's footsteps, the current Attorney General of the United States, Eric Holder, declared the United States to be 'a nation at war' (quoted in Yoo, 2012). Holder stated in March 2012 that 'in this war, we face a nimble and determined enemy that cannot be underestimated' (see Holder, 2012). This war rhetoric was further intensified by Holder, using statements that justified adopting *all necessary measures* to protect the American people. As Holder has argued 'disrupting and preventing these plots – and using every available and appropriate tool to keep the American people safe – has been, and will remain, this Administration's top priority' (ibid.). Identifying its targets as 'enemy belligerents' Holder argued that the United States was 'in an armed conflict' and it was authorised to take action against its enemies under international law (ibid.).

The similarities between the two Administrations are at their starkest when the Obama Administration makes the case for acting pre-emptively against its enemies using drones. The Bush administration's well known *National Security Strategy* of 2002 highlighted a vision for America to act pre-emptively when an attack was imminent (National Security Strategy, 2002). This view led to the American attack on Iraq which took place before, it was claimed, Saddam Hussain could attack the US with his purported chemical and biological weapons (Slocombe, 2003). The current White House Administration has also sought to justify the use of drones as pre-emptive actions against those planning to attack America. Obama's counterterrorism chief John Brennan said that any lethal action against terrorists through drones is not 'about punishing them for their past crimes' because the United States is 'not seeking vengeance' (Brennan, 2012). The main reason for strikes however is 'to mitigate an actual ongoing threat, *to stop plots, prevent future attacks, and to save American lives*' (ibid., emphasis added). Attorney General Holder has argued that 'the Constitution empowers the President to protect the nation from any *imminent* threat of violent attack. And international law recognises the inherent right of national self-defense' (Holder, 2012). This argument was also employed by the Bush Administration in its approach to tackling terrorism, highlighting the continuities between the two administrations. Such a use of a 'war paradigm' is essential to gain acceptance and legitimacy for potentially unpopular American policies in the eyes of American public who need to be persuaded of the severity of the threat necessitating them.

Washington officials have also been trying to blur the differences between various terrorist groups operating within the tribal areas of Pakistan by defining them as 'Al-Qaeda and associated forces' (Holder, 2012). According to the Obama Administration, it is also entitled to go after the affiliates of Al-Qaeda because they can be used by the group to conduct anti-US activities when it is too weak to do so itself. For example, Brennan stated that 'Al-Qaeda's leadership ranks have continued to suffer heavy losses' (Brennan, 2012). He went on to say that 'as the Al-Qaeda core falters, it continues to look to *affiliates and adherents* to carry on its murderous cause' (ibid., emphasis added). However, this cannot be seen as a casual shifting the focus towards random 'affiliates' and 'adherents' of Al-Qaeda. It can only be seen as an attempt to gain legitimacy for conducting drone attacks on almost anyone in the tribal areas of Pakistan whom the Administration could merely pronounce as an 'affiliate' of Al-Qaeda. This is the phenomenon of threat construction highlighted in the earlier part of this chapter, i.e. the reality as well as threats around us are created by human actors through their actions and they don't just exist 'out there' in an objective form. The officials in the US have asked the public to trust them to decide and define who the external enemy is, in the context that they are fighting a 'war' on behalf of the country in which its very existence is in danger. The audience is given very little choice except to trust the securitising actor (John Brennan in this case) and hope that its government's policies will help alleviate the threat.

The Bush White House justified its war on terrorism under the Authorization of the Use of Military Force (AUMF) that was granted by the Congress to the Administration on 14 September 2001 (Resolution, 2001). The current Administration has also justified its drone attacks on the legal basis of the same AUMF (Koh, 2010). Though the AUMF resolution was passed with Afghanistan in mind and is more than a decade old, the current White House has stretched the directive's territorial limits to make it applicable to its action within Pakistan as well. There is some weight in arguing that the United States cannot just focus on Afghanistan to protect itself when the threat from Al-Qaeda may be emanating from the neighbouring country that may be either unwilling or unable to take action against terrorists on its territory (Ralph, 2010). Ralph, however, argues that such an approach could be convincing only 'to the extent that one sees Al-Qaeda in homogenous terms and sees all Islamist inspired attacks as somehow directed by the Al-Qaeda leadership' (ibid.). That clearly is not the case, however, as the drones have targeted not only Al-Qaeda militants but other individuals unrelated to Al-Qaeda; indeed, targets may be members of the Taliban Movement of Pakistan (TTP), the Afghan Taliban or other, unrelated groups.[8] The Obama Administration not only first forcefully put forward the tribal region as a legitimate area of its activity but then also brought disparate groups within the umbrella of Al-Qaeda (knowing the political impact on its home audience) specifically to target them. This was done using phrases like 'Al-Qaeda affiliates'. One of the key 'produced absences' is simple absence of a detail of differences between various militant groups (Hopf, 1998: 176). In brief, US policymakers constructed a threat and the officials bestowed upon themselves a legitimate right to act against that threat. Their portrayal of other groups as being part of that threat is convenient to their message and overall goal, and inconvenient details are not allowed to get in the way.

Mirroring the Bush-era rhetoric, Holder (2012) said that

> military and civilian officials must often make real-time decisions that balance the need to act, the existence of alternative options, the possibility of collateral damage, and other judgments … all of which depend on expertise and immediate access to information that only the executive branch may possess in real time.

This is similar to saying that the American people can trust the Obama Administration because it is the only entity that knows the true extent of the threat and how to deal with it. This rhetoric is employed to justify secrecy in the decision making processes as well. The Obama Administration's insistence that only it had the right to decide who actually *is* Al-Qaeda is replete with legal and ethical ambiguities. The American Civil Liberties Union (a vocal opponent of the use of drones) has repeatedly argued against the power being left in the hands of the US government 'to kill people anywhere in the word, including citizens, on the basis of legal standards and evidence that are never submitted to the court' (quoted in BBC News, 2012).

The current Administration also argued that a key difference between itself and its predecessor was that the Bush White House declared a war against all terrorists wherever they existed (Ralph, 2010). However, in his major May 2012 speech to justify the use of drones, John Brennan pointed to potential targets in Pakistan, Yemen, Somalia, Nigeria, and elsewhere in West Africa (see Shah, 2012). Quite like the Bush White House, the current Administration has favoured pre-emptive action against whomever it considered an enemy, in the form of Al-Qaeda or its affiliates. The action to eliminate this threat would of course be taken once the Administration (and nobody else) decided that an actor posed a threat to American lives.

Framing its activity in this way has had significant implications, as doing so entitled Washington to argue that the US continues to remain at war with Al-Qaeda – a politically very convenient stance. The US has granted itself the ability to shoot and kill its enemies (using drones, in this case) who would normally expect to be arrested and prosecuted under a simple law-enforcement mechanism (Roth, 2004; see also Savage, 2010). The 'law-enforcement' rules are codified in international human rights law that afford many more rights to individuals than the laws of war would (Roth, 2004). This distinction has had significant implications for the trials of 'unlawful combatants' in the military court as well as the continuing detention of detainees in the US facilities at Guantanamo Bay (see Schulhofer, 2009).[9] The Bush Administration defined its enemies as 'enemy combatants' and then detained them without charge rather than prosecute them – as doing so was a 'routine and well-accepted' practice during times of war (Hafetz, 2012: 845). The continuous usage of the war paradigm has, however, had an impact on global audiences. 'The notion of a global armed conflict against Al-Qaeda and associated terrorist organizations took root, gaining acceptance among judges, lawmakers, and the public' (Hafetz, 2012: 848), thus completing the process through which a threat is constructed and one attempts to gain a worldwide legitimacy for an action against that threat.

Conclusion

This chapter has examined the political statements of the two actors who in turn oppose and support the use of drones in Pakistan's tribal areas. It argues that the opposition to drones by Imran Khan serves a clear political purpose whereby he tries to discredit the country's traditional rulers and portrays himself as the alternative to those ruling elites. Khan employs his rhetoric to gain legitimacy in the eyes of the masses. Given that Khan's party was able recently to gain power in the northern province of KPK (near the affected areas) this indicates that this rhetoric may indeed have helped him win power. It also indicates a successful securitisation of the issue by Khan, as the process is only complete when the audience accepts the threat as such. During his campaign, Khan asserted that if the PTI were to form the next national government in Pakistan, it would shoot down American drones – a nakedly populist stance (Khan, 2012a). He stated: 'we will not allow the United States to conduct its drone attacks in Pakistan. If the United States continued its attacks, then we will shoot down their drones and we will protest in the United Nations' (Khan, 2012a). The coming years will reveal how Khan's regional government in the KPK opposes drones. It will be useful to see whether Khan and his party choose to employ the anti-drone rhetoric to divert attention from other major challenges the KPK may be facing – including poverty, illiteracy, and lack of healthcare.

In the US, the Administration of Barack Obama has continued to justify the use of drones as part of America's ongoing war against terrorism. A close look at Obama Administration officials' speeches and statements on the issue of drones highlights that a 'war paradigm' is continuously created and maintained to justify the use of unmanned aerial vehicles. As it provides him considerable political capital, such a policy is likely to continue during President Obama's second term in office.

Notes

1 The author would like to thank Safi Aslam for his help with the proofreading of this chapter.
2 For the current author's work on drones see Aslam (2011, 2012).
3 It would be useful to study how threats have been socially constructed in other theatres such as Somalia, Afghanistan, and Yemen. However, such a study is not feasible given a serious dearth of primary and secondary literature concerning those theatres.
4 The PML (N) is currently in power in Pakistan led by the country's current Prime Minister, Nawaz Sharif.
5 General Pervez Musharraf served as the country's leader from 12 October 1999 until 18 August 2008.
6 Asif Ali Zardari is the former president of Pakistan. He assumed the office of President on 9 September 2008.
7 This chapter was written before the current offensive in North Waziristan commenced.
8 Afghan Taliban are usually seen as more closely aligned with Al-Qaeda while

there have been differences between Afghan Taliban and Pakistani Taliban. See *The Indian Express* (2012).

9 On the issue of the link between detention/trial of the enemy combatants and the 'global war' paradigm, see Schulhofer (2009).

References

Abouzeid, R. (2010) 'Pakistan's Military Holds Back in North Waziristan', *Time* (17 April 2010).

Aslam, M. W. (2011) 'A Critical Evaluation of American Drone Strikes in Pakistan: Legality, Legitimacy and Prudence', *Critical Studies on Terrorism*, 4, 3: 313–29.

Aslam, M. W. (2012) 'Understanding the "Pak" in "AfPak": The Obama Administration's Security Policy for Pakistan at the Mid-Term', *Journal of Policing, Intelligence and Counter Terrorism*, 7, 1: 2–21.

Balzacq, T. (2005) 'The Three Faces of Securitization: Political Agency, Audience and Context', *European Journal of International Relations*, 11, 2: 171–201.

BBC News (2012) 'White House in First Detailed Comments on Drone Strikes' (1 May 2012).

Becker, J. and Shane, S. (2012) 'Secret "Kill List" Proves a Test of Obama's Principles and Will', *New York Times* (29 May 2012).

Blank, L. (2011) 'The Consequences of a "War" Paradigm for Counterterrorism: What Impact on Basic Rights and Values?' *Georgia Law Review*, 46: 719–41.

Brennan, J. (2012) 'The Efficacy and Ethics of U.S. Counterterrorism Strategy', *Transcript of Remarks by John O. Brennan, Assistant to the President for Homeland Security and Counterterrorism, Wilson Centre, Washington DC*, 30 May, www.wilsoncenter.org/event/the-efficacy-and-ethics-us-counterterrorism-strategy (accessed 4 March 2013).

Bukhari, M. (2012) 'Pakistan Court Rules Army Must Stop Interfering in Politics', *Reuters* (19 October 2012).

Burke, J. (2012a) 'Imran Khan: Set to Play the Innings of His Life', *Guardian* (1 January 2012).

Burke, J. (2012b) 'Imran Khan: The Man Who Would be Pakistan's Next Prime Minister', *Guardian* (4 March 2012).

Burkeman, O. (2009) 'Obama Administration Says Goodbye to "War On Terror"', *Guardian* (25 March 2009).

Bush, G. (2001) 'President Bush's Address to the Congress', *BBC News* (21 September 2001).

Buzan, B., Wæver, O., and Wilde, J.D. (1998) *Security: A New Framework for Analysis*, Boulder, CO: Lynne Rienner.

Channel 4 (2012) 'Drones Will Push People towards Terrorism' (24 September 2012).

Cheney, D. (2003) 'Interview, Former US Vice President Dick Cheney', *Meet the Press*, NBC television broadcast (14 September 2003), www.msnbc.msn.com/id/3080244/ns/meet_the_press/t/transcript-sept (accessed 13 March 2013).

Cockburn, P. (2010) 'Pakistan Won't Act Against the Taliban', *Independent* (2 October 2010).

Collins, A. (2005) 'Securitization, Frankenstein's Monster and Malaysian Education', *The Pacific Review*, 18, 4: 567–88.

Crilly, R. (2011) 'Imran Khan: I Have the Opposition on the Mat – My Time Has Come', *The Telegraph* (25 December 2011).

Daily Times (2011) 'Imran Khan Declares Legal War against Drone Attacks' (20 April 2011).

Dawn (2013) 'PTI Won Second Most Votes in Pakistan Poll' (27 May 2013).

Dotty, R.L. (1993) 'Foreign Policy as Social Construction: A Post-Positivist Analysis of U.S. Counterinsurgency Policy in Philippines', *International Studies Quarterly*, 37, 3: 297–320.

Emmers, R. (2007) 'Securitization', in Collins, A. (ed.) *Contemporary Security Studies*, Oxford: Oxford University Press, pp. 136–51.

Enemark, C. (2011) 'Drones over Pakistan: Secrecy, Ethics and Counterinsurgency', *Asian Security*, 7, 3: 218–37.

Fierke, K. (2007) *Critical Approaches to International Security*, Oxford: Oxford University Press.

Financial Times (2013) 'Imran Khan Enjoys Late Surge in Pakistan Polls' (10 May 2013).

Finnemore, M. (1996) *National Interests in International Society*, Ithaca, NY: Cornell University Press.

Hafetz, J. (2012) 'Redefining State Power and Individual Rights in the War on Terrorism', *Valparaiso University Law Review*, 46: 843–57.

Holder, E. (2012) 'Attorney General Eric Holder Speaks at Northwestern University School of Law', United States Departement of Justice, 5 March, www.justice.gov/iso/opa/ag/speeches/2012/ag-speech-1203051.html (accessed 4 March 2013).

Hopf, T. (1998) 'The Promise of Constructivism in International Relations Theory', *International Security*, 23, 1: 171–200.

Jefferies, S. (2011) 'Imran Khan: America is Destroying Pakistan. We're Using Our Army to Kill Our Own People with Their Money', *Guardian* (18 September 2011).

Khan, I. (2008) 'Petraeus Asked to Try The Political Approach', *Dawn* (5 November 2008).

Khan, I. (2011) 'Interview: Imran Khan', *CNN* (15 April 2011), www.youtube.com/watch?v=193lW2-RQ-g (accessed 15 March 2013).

Khan, I. (2012a) 'Interview: Imran Khan', *Geo TV Programme Lekin* (16 April 2012), www.youtube.com/watch?v=99kK0eFGo2g (accessed 15 March 2013).

Khan, I. (2012b) *Pakistan: A Personal History*, London: Bantam.

Koh, H. (2010) 'The Obama Administration and International Law', *US Department of State* (25 March 2010), www.state.gov/s/l/releases/remarks/139119.htm (accessed 4 March 2013).

MacDonald, M. (2002) 'Human Security and the Construction of Security', *Global Society*, 16, 3: 277–295.

MacDonald, M. (2005) 'Constructing Insecurity: Australian Security Discourse and Policy Post-2001', *International Relations*, 19, 3: 297–320.

MacDonald, M. (2008) 'Securitization and the Construction of Security', *European Journal of International Relations*, 14, 4: 563–87.

MacDonald, M. and Merefield, M. (2010) 'How Was Howard's War Possible? Winning the War on Position over Iraq', *Australian Journal of International Relations*, 64, 2: 186–204.

Masood, S. (2013) 'Pakistan Cricket Star Tries Again to Turn Adulation into Political Support', *New York Times* (23 March 2013).

National Security Strategy (2002) *The United States National Security Strategy of 2002*, http://georgewbush-whitehouse.archives.gov/nsc/nss/2002 (accessed 4 March 2013).

Nawaz, S. (2009) *Crossed Swords: Pakistan, its Army and the Wars Within*, Karachi: Oxford University Press.

Nelson, D. (2011) 'Imran Khan Bowls Back onto Pakistan Political Agenda with Mass Lahore Rally', *The Telegraph* (31 October 2011).

Ralph, J. (2010) 'Continuity We Have To Believe In (Part II)', *Law, War and the State of American Exception Blog* (30 March 2010), http://lawwarandthestateoftheamericanexception.wordpress.com/2010/03/30/continuity-we-have-to-believe-in-part-ii (accessed 15 April 2013).

Resolution (2001) Public Law 107–40 (18 September 2001), www.gpo.gov/fdsys/pkg/PLAW-107publ40/pdf/PLAW-107publ40.pdf (accessed 4 March 2013).

Ross, A.K. (2013) 'Incoming CIA Boss Says Covert Drone Strikes Are "Last Resort"', *The Bureau of Investigative Journalism* (8 February 2013), www.thebureauinvestigates.com/2013/02/08/incoming-cia-boss-says-covert-drone-strikes-are-last-resort (accessed 4 March 2013).

Roth, K. (2004) 'Drawing the Line: War Rules and Law Enforcement Rules in the Fight against Terrorism', *Human Rights Watch World Report 2004*, January, www.globalissues.org/article/469/war-rules-and-law-enforcement-rules-in-terror-war (accessed 4 March 2013).

Savage, C. (2010) 'Obama Team is Divided on Anti-Terror Tactics', *New York Times* (28 March 2010).

Schulhofer, S. (2009) *Unravelling Guantanamo: Detentions, Trials and the 'Global War' Paradigm*, Public Law and Legal Theory Research Paper Series, Working paper Number 09–08, New York University School of Law.

Shah, N. (2012) 'Drone Attacks and the Brennan Doctrine', *Guardian* (2 May 2012).

Slocombe, W. (2003) 'Force, Pre-emption and Legitimacy', *Survival*, 45, 1: 117–30.

Taj, F. (2010) 'The Year of the Drone Misinformation', *Small Wars and Insurgencies*, 21, 3: 529–35.

Taqi, M. (2010) 'Imran Khan: A Taliban Goebbels', *Daily Times* (5 August 2010).

The Express Tribune (2012) 'Honour and Shame' (18 April 2012).

The Express Tribune (2011) 'PML-N Warns Army, ISI to Stop Supporting PTI' (13 November 2011).

The Indian Express (2012) 'Mullah Omar Tells TTP to Stop Attacking Pakistan Army' (3 January 2012).

The Nation (2013) 'Destroyers Cannot Be Constructors, Says Imran' (6 May 2013).

The News (2013a) 'Imran Vows to Shoot Down US Drones', *The News Pakistan*, (4 March 2013).

The News (2013b) 'Imran Khan Attacks on Polling Offices', *The News Pakistan* (26 April 2013).

The Times of India (2011) 'Gilani Criticises United States for Fresh Drone Strikes' (14 April 2011).

Wæver, O. (1995) 'Securitization and Desecuritization', in Lipschutz, R.D. (ed.) *On Security*, New York: Columbia University Press, pp. 46–86.

Woods, C. (2012a) 'Obama Outs CIA Drone Campaign – but do His Words Add up?', *The Bureau of Investigative Journalism* (1 February 2012), www.thebureauinvestigates.com/2012/02/01/analysis-obama-outs-secret-cia-drone-campaign-but-do-his-words-add-up (accessed 2 March 2013).

Woods, C. (2012b) 'Evidence in British Court Contradicts CIA Drone Claims', *The Bureau of Investigative Journalism* (24 April 2012), www.thebureauinvestigates.com/2012/04/24/british-legal-case-shines-fesh-light-on-civilian-drone-deaths (accessed 4 March 2013).

Yoo, J. (2012) 'The Good and Bad in Eric Holder's Drone Defense', *The Wall Street Journal* (12 March 2012).

Yusufzai, R. (2012) 'The Political Drones Get Louder', *Inter Press Service* (17 May 2012).

Part II

Legal and moral implications

5 Covert drone strikes and the fiction of zero civilian casualties[1]

Chris Woods

Introduction

The US Central Intelligence Agency and Pentagon have killed an estimated 2,800 or more people in covert drone strikes in Pakistan, Yemen, and Somalia.[2] These attacks are cited as a key factor in the decimation of 'central' Al-Qaeda and other militant organisations. The CIA asserts that civilian deaths are kept to a minimum and that no non-combatant was killed in Pakistan during a 16-month period of intense bombing.[3] There is compelling evidence to suggest otherwise, indicating a need for significantly more transparency on the part of the US government on casualty data.

The perfect weapon?

In March 2013 Barack Obama publicly confirmed that civilians had died as a consequence of the US's covert drone war – the targeted killing of hundreds of individuals outside of the 'hot' battlefield. 'It is a hard fact that U.S. strikes have resulted in civilian casualties, a risk that exists in every war. And for the families of those civilians, no words or legal construct can justify their loss', the President told an audience at the National Defense University in Washington, D.C. (Obama, 2013). Without appearing to take a position himself, Obama also acknowledged the gulf in understanding between public and US government estimates on this contentious issue, noting: 'Much of the criticism about drone strikes – both here at home and abroad – understandably centres on reports of civilian casualties. There's a wide gap between U.S. assessments of such casualties and nongovernmental reports.' Obama's frank comments were a welcome step forward. Prior to his speech US officials had tended to play down or even deny credible claims of non-combatant deaths in countries such as Pakistan and Yemen; for example, Leon Panetta, US Defense Secretary, told PBS: 'I have to tell you as [former] director of the CIA – and that's been true not only for me but for those that have followed me – that if there are any civilians in the shot, you don't take it' (Panetta, 2011).

Indeed, for a significant period the US intelligence community sought to portray an unfeasible reality – one in which armed drones had reached a point of such perfection that they were no longer killing civilians at all.

Remotely piloted aircraft (RPAs) – more generally referred to as armed drones – have emerged as a significant tool of US counter-terrorism and counter-insurgency operations since 2001. The majority of US drone strikes have been carried out in conventional theatres of war such as Iraq, Afghanistan, and Libya, where their use is overtly governed by international humanitarian law. In the event of reported civilian casualties, public or semi-public accountability mechanisms are generally in place. Inquiries can be initiated; their findings publicly reported; and where required, compensation paid to victims' families (Garlasco, 2012).

In a parallel (and often related) bombing campaign, drones have also been used extensively by the United States since 2002 to carry out targeted killings outside of the conventional war theatre. Here, separate rules appear to be in play. Whilst specific details remain classified, there are credible indications that more than 400 US covert drone strikes had, by November 2013, killed a minimum of 2,800 people, most of whom are claimed to have been low-level militants.[4] The majority of these attacks took place in Pakistan and Yemen, with a small number in Somalia. The strikes have mainly been the work of the Central Intelligence Agency, which carries out all bombings in Pakistan and many of those in Yemen. The Pentagon's Joint Special Operations Command (JSOC) – part of US Special Operations Command (USSOCOM) – is responsible for all drone strikes in Somalia, and also operates an independent drone fleet above Yemen. Until 2012 the entire US covert drone campaign was considered a state secret, and much effort was exerted to protect it. Journalists, academics, and legal campaigners had failed to convince US federal courts that key aspects of the programme should be revealed. In December 2010, for example, a US Federal Court dismissed an attempt by Nasser al-Awlaki to halt the targeting of his US-born son by the Central Intelligence Agency (*Al Aulaki v. Obama et al.*, 2010). However pronouncements by senior members of the Obama administration have led to a de facto partial declassification of the campaign. The President himself first directly referred to CIA drone strikes in Pakistan's tribal areas in a Google 'Town Hall' debate. Admitting that 'a lot of these strikes have been in the FATA [Federally Administered Tribal Area], and going after al Qaeda suspects', Obama also insisted that 'drones have not caused a huge number of civilian casualties' (Obama, 2012a). In June 2012 the president informed Congress that the US was carrying out 'direct military action' in Yemen and Somalia against Al-Qaeda and its affiliates – partially declassifying the US drone campaign in these countries (Obama, 2012b). Attorney General Eric Holder, chief counter-terrorism adviser, and later CIA director John Brennan, and the general counsels of the Pentagon, CIA, and State Department have all spoken at some length – and on the record – about the US targeted killing strategy (see, for example, Holder, 2012 and Brennan, 2012b). Anonymous senior intelligence figures have also regularly briefed journalists and academics about the on-going programme, often revealing significant details. In May 2012 for example, 36 present and former US officials co-operated with a major *New York Times* investigation into the US covert drone campaign (Becker and Shane, 2012). Missing from this welcome discourse, however, has been any comprehensive detail on who the United States believed it was killing with drones – and how many of those it understood to be civilians. These would appear to be some of the most closely guarded secrets of the campaign.

Civilian casualties in Pakistan: the US position

Although in the earlier years of the covert campaign US officials occasionally and anonymously commented on individual strikes, overall US estimates of civilian casualties in Pakistan did not begin to emerge until late 2009. Media organisation *ProPublica* collated all estimates of civilian casualties in Pakistan, as iterated by (generally anonymous) US officials (Elliott *et al.*, 2012). The first such claim appeared in the *New York Times.* Referring to the period between January 2007 and December 2009, an anonymous government official told the newspaper: 'We believe the number of civilian casualties is just over 20, and those were people who were either at the side of major terrorists or were at facilities used by terrorists' (Shane 2009). Since then US officials have offered estimates of civilian casualties in Pakistan on approximately 25 occasions. Generally the US has insisted that since 2004 it has killed no more than 50 to 60 civilians in Pakistan, although there are some contradictory reports. One claim – which first appeared in May 2012 – was that civilians 'typically in the single digits' had been killed in Pakistan since Obama came to office (Becker and Shane, 2012). Yet in April 2010 the *Wall Street Journal* (Johnson 2010) and the *Washington Post* (Warwick and Finn, 2010) both reported US officials as saying that an estimated 20 civilians had already died under Obama. The President himself, speaking in January 2012, asserted that 'drones have not caused a huge number of civilian casualties, for the most part they have been very precise precision strikes against al Qaeda and their affiliates' (Obama, 2012a). An overall estimate of those killed by the CIA in Pakistan was offered for the first time in August 2012, when anonymous US officials told the *New York Times* and others, this author included, that alongside approximately 50 non-combatants killed since 2001 were an estimated 2,000 militants[5] (*New York Times*, 2011).

Having firmly established a position that no more than 50–60 civilians had died in more than six years of Pakistan drone bombings, in late January 2011 some US officials began asserting a more aggressive position. News agency Bloomberg first cited an unnamed US government official as saying that 'since the drone program accelerated in mid-August [2010 we] have killed several hundred militants without causing any deaths among civilian non-combatants' (Capaccio and Bliss, 2011). Three weeks later, the *Los Angeles Times* identified a drone strike on 23 August 2010 as being the last occasion on which civilians had died.[6] A Pakistani official told the newspaper that this reflected a decision by the US to use greater caution in its drone attacks. A US official, too, stated that in more than 75 strikes in Pakistan's tribal areas since August 2010, no civilian had been killed, 'suggesting that the CIA, which runs the drone program in Pakistan, has been more judicious in its attacks' (Dilanian, 2011). On 29 June 2011, President Obama's then-chief counter-terrorism adviser John Brennan became the first senior, named, official in the Administration to proclaim publicly that civilians were no longer dying in US drone strikes. Speaking at Johns Hopkins University, Brennan noted in a question and answer session:

> One of the things President Obama has been insistent upon is that we're exceptionally precise and surgical in terms of addressing the terrorist threat. And by that I mean, if there are terrorists who are within an area where there are women and children or others, you know, we do not take such action that might put those innocent men, women, and children in danger.

Without referring directly to the 23 August 2010 CIA attack, Brennan added that for almost a year 'there hasn't been a single collateral death because of the exceptional proficiency, precision of the capabilities that we've been able to develop' (Brennan, 2011). In subsequent weeks, US counter terrorism officials confirmed to this author and other journalists that the CIA shared Brennan's assessment – that civilians were no longer being killed.[7]

The fiction of zero civilian casualties

John Brennan's public claims of zero civilian deaths in Pakistan over a lengthy period were not at first contested by academics or the media. As a national security correspondent of one major US newspaper put it, 'We probably took our eye off the ball on this one.'[8] The first overt challenge came from the London-based Bureau of Investigative Journalism (BIJ, the 'Bureau').[9] Close to completing a major reassessment of all known CIA drone strikes in Pakistan as part of a new meta-database, the BIJ appeared well positioned to test claims of zero civilian casualties. UK-based researchers first identified some 116 US drone strikes which had taken place in Pakistan between 23 August 2010 and 29 June 2011, the date of Brennan's speech (which were estimated to have killed 700 or more people in total). Claims of civilian casualties by credible media were identified in approximately one quarter of these strikes. The Bureau then worked with reputable Pakistani-based journalists to interrogate these claims. Three field researchers – two in North Waziristan and one in South Waziristan – sought to confirm locally whether civilians had died, and to name them where possible. As the Bureau's July 2011 report concluded:

> While we cannot always be categorical, in at least 10 of the 25 cases we have identified, we understand that civilians were killed. The evidence shows that at least 45 civilians died in these strikes. Six of them were children under 16 years old.[10]

Fifteen additional incidents 'raised significant questions. In these cases at least 65 civilian deaths have also been reported but are contested, or are reported by a single source only' (Woods, 2011). Despite the Bureau's assertions, the US intelligence community continued to stand by its claim that no civilians had died in the stated period. Speaking on background terms in mid-July 2011 to this author, a US counter-terrorism official insisted:

> There have been no non-combatant casualties as a result of US counter-terrorism operations in Pakistan since last August. Extraordinary care is taken in the course of these operations to ensure precision and to focus on the protection of non-combatants who may be nearby. This is the most precise weapon in the history of warfare.

Later again, US officials extended the no-deaths period, claiming that zero civilians had died since May 2010 (Shane, 2011a).

Contradictory evidence

The following year, John Brennan sought to clarify his own 29 June 2011 comments, telling George Stephanopoulos that 'what I said was that over a period of time before my public remarks that we had no information about a single civilian, a non-combatant being killed' (Brennan, 2012a). Neither of Brennan's statements appears to bear scrutiny.

On the morning of 17 March 2011, missiles from US drones struck a gathering of several dozen men in the village of Datta Khel in North Waziristan. According to survivor testimony, the damage was so severe that the adjacent bus station and buildings were still burning some hours later (Woods, 2012b).[11] As many as 40 people died in the attack. Anonymous US officials immediately claimed that those killed were Taliban and Al-Qaeda terrorists. One told the *New York Times*: 'These people weren't gathering for a bake sale. They were terrorists' (Masood and Shah, 2011). The *Wall Street Journal* also reported an anonymous US official – possibly the same individual – as saying: 'These guys were terrorists, not the local men's glee club' (Wright and Mehsud, 2011). However, numerous field investigations have since concluded that those killed were almost all civilians, gathered for the second day of a tribal *jirga* or elders' meeting to discuss an on-going dispute over mining rights in the area. Brigadier Abdullah Dogar, who was then Pakistan's military commander in Waziristan, later told the *Sunday Times*:

> We in the Pakistan military knew about the meeting, we'd got the request ten days earlier. It was held in broad daylight, people were sitting out in Nomada bus depot when the missile strikes came. Maybe there were one or two Taliban at that *Jirga* – they have their people attending – but does that justify a drone strike which kills 42 mostly innocent people?
>
> (Lamb *et al.*, 2012)

Many of those killed were *maliks* – tribal elders critical to local peacekeeping efforts. One such man was Ismail Khan. In a sworn affidavit presented in the London High Court in 2012, Khan's son stated: '[My father] always did the right thing for the community and the tribe. He opposed terrorism and militancy and was not himself in any way connected to these things' (Woods, 2012b).

The deaths of so many civilians, including tribal elders, generated an unprecedented storm of criticism in Pakistan. A significant civilian death toll was widely reported by local and international media. General Ashfaq Kayani, head of the Pakistan military (with whom the US government had extensive connections), issued a unique public criticism of the drone strikes, in which the general stated that he

> strongly condemns the Predator Strike carried out today in North Waziristan Agency resulting into loss of innocent lives. It is highly regrettable that a *jirga* of peaceful citizens including elders of the area was carelessly and callously targeted with complete disregard to human life.
>
> (Kayani, 2011)

President Asif Ali Zardari told a visiting US Congressional delegation that the attack was 'intolerable' (Dawn, 2011b). And Prime Minister Yousuf Raza Gilani stated that 'such an irrational behaviour negatively impacts efforts to separate militants from peaceful and patriotic tribesmen of the areas' (Dawn, 2011a). Cameron Munter, the United States ambassador to Pakistan at the time, was summoned on 19 March 2011 to the Ministry of Foreign Affairs to receive Islamabad's protests. As the official Ministry press release noted that day:

> Ambassador Munter was categorically conveyed that such strikes were not only 'unacceptable' but also constituted 'a flagrant violation of humanitarian norms and law' ... Ambassador Munter said that he

> understood clearly that this was not a *proforma* demarche. He will rush to Washington to convey Pakistan's message to the US Administration at the highest levels.
>
> (Pakistan Ministry of Foreign Affairs, 2011)

A former US official with knowledge of the meeting has confirmed to this author that Pakistan's protest at the significant loss of civilian life was indeed conveyed to 'the highest levels' of the Obama administration. There is no doubt, then, that Washington was formally apprised of multiple civilian deaths in the 17 March 2011 strike. It is also inconceivable that the US intelligence community was unaware of these reports. Yet 14 weeks later, President Obama's chief counter-terrorism adviser John Brennan would claim that no 'non-combatants' had been killed in Pakistan since August 2010 (Brennan, 2011). And as noted, he would also later insist that 'that over a period of time before my public remarks [that] we had no information about a single civilian, a non-combatant being killed' (Brennan, 2012a).

The CIA's definition of 'non-combatant'

As the 17 March *jirga* incident demonstrates, there was ample evidence available to the United States within hours of the attack that non-combatants had died. It is also clear from Brennan's statements that such casualties were not recorded as civilian deaths by the US government, a position it continues to maintain. In August 2011, for example, an anonymous US counter intelligence official told the *New York Times*:

> There's no question the Pakistani and US governments have different views on the outcome of this strike. The fact is that a large group of heavily armed men, some of whom were clearly connected to Al Qaeda and all of whom acted in a manner consistent with A.Q.-linked militants, were killed.
>
> (Shane, 2011b)

In February 2012 Associated Press noted that US officials continued to maintain this position. Yet that same AP investigation – based on eyewitness reports from Datta Khel – itself found that a large number of civilians had died in the town (Abbott, 2012). Similar field studies by Stanford and NYU law schools, by the BIJ, and by legal charity Reprieve have all reached similar conclusions, as did an internal estimate by the FATA Secretariat. How is it then that the CIA refuses to acknowledge civilian deaths in this incident?

Public casualty estimates by organisations such as the New America Foundation, BIJ, and the Long War Journal are drawn predominantly from media reports. These in turn are based on interviews with local residents, with Pakistani (and occasionally US) officials, and/or with Taliban representatives. Although concern has been expressed periodically regarding the accuracy of such reporting, recent studies indicate a reasonable level of accuracy. The extensive field investigation by Associated Press published in February 2012 examined the 10 Pakistan drone strikes with the highest-reported casualty count since January 2009. Almost 80 civilian eyewitnesses were interviewed for the study. AP reported a close correlation between the *overall* casualty number initially reported by local officials, and its own subsequent findings via field work. However, as AP noted, 'the main difference was that the intelligence officials often did not distinguish [in their casualty reports] between militants and civilians'[12] (Abbott, 2012). There actually appears little difference between the CIA's overall drone death tally and the baseline estimates of monitoring bodies. In August 2011 the CIA reported that it had killed some 2,050 individuals in its Pakistan drone strikes. The Long War Journal's own estimates for the period are within 5 per cent of this figure (2,152 killed), with BIJ reporting 2,135 deaths – some 4 per cent higher than the CIA's tally.[13] Yet all monitoring organisations also show a far higher reported civilian death toll than the US admits to. What remains at issue then is not how many are being killed, but how they are then categorized. Yet efforts to seek clarification from the CIA regarding its definition of a 'non-combatant' have been consistently rebuffed. Agency coyness regarding such queries may simply relate to the highly classified status of the campaign. However, critics have argued that this may also disguise the fact that the CIA employs definitions of non-combatants which would not be recognised by conventional military forces.

In May 2012 the *New York Times* gave a valuable insight into why US government estimates of civilian deaths may be so at odds with public counts. In an extensive feature based on interviews with three dozen present and former US officials, the *Times* reported:

> Mr. Obama [has] embraced a disputed method for counting civilian casualties that did little to box him in. It in effect counts all military-age males in a strike zone as combatants, according to several administration officials, unless there is explicit intelligence posthumously proving them innocent.
>
> (Becker and Shane, 2012)

No US Administration figure has subsequently challenged this representation. By labelling all military-age males as 'combatants', the US Administration has likely been able radically to reduce internal estimates of civilian drone strike casualties. Such activity may not be restricted to Pakistan. A former intelligence analyst attached to US Special Operations Forces (SOF) in Afghanistan, informed this author that it was his experience throughout 2012 that 'military-aged males' of indeterminate status were routinely killed in Afghan SOF drone strikes on the basis of 'guilt by association' with nearby targets of interest. Only the observed presence of women and children would normally trigger concerns regarding civilian casualties.[14]

All of those civilian males killed at the 17 March 2011 *jirga* are likely to be 'combatants', in the view of the United States, simply by dint of their age and gender. Moreover, despite claims in the *New York Times* that fatalities may be reclassified given 'explicit intelligence posthumously proving them innocent' there is no evidence that this takes place. Despite a number of credible field investigations confirming a high number of civilian deaths in the 2011 Datta Khel strike and other incidents since August 2010, US civilian estimates have not been adjusted, remaining stubbornly at 50 to 60 deaths (Ferran, 2013).

The Datta Khel incident was one of a number of US drone strikes between August 2010 and June 2011 in which civilian deaths were well reported. Just two weeks prior to John Brennan's speech, five men were killed in their vehicle by a CIA remotely piloted aircraft in the North Waziristan town of Mir Ali. Although the deceased were initially reported to be militants, an alternative narrative swiftly emerged. According to villagers, none of the men was aligned with any terrorist or militant group. The dead were named as Shehzad, a student; his cousin Akram Shah, a government employee who had been driving the car; Atiq ur Rehman, nicknamed Tariq, a local pharmacist; Irshad Khan, who worked for Mr Rehman; and Amar Khan, a student at Miranshah college. A field researcher for the BIJ later noted that one of those killed was the grandson of anti-Taliban tribal elder Malik Shahzada, the 50-year-old head of the Toorikhel Wazir tribe. Shahzada had even fled to Peshawar with most of his family to avoid Taliban attacks. In such circumstances, 'how could he or his slain grandson be Taliban?' the researcher queried (Woods and Yusufzai, 2011).

A field study by two major US universities also reported that five civilian males were killed in the attack, following detailed interviews with villagers and family members (Stanford and NYU, 2012). Villagers not only publicly protested the deaths of the men but blocked a major road with their coffins on the day of the funeral, insisting that all were civilians (*The News International*, 2011). This was a highly unusual move. Potential signifiers of non-combatant deaths in Waziristan can include public protests over civilian casualties, and the burying of victims in local graveyards. Neither is generally observed when Taliban or other militants are killed. Both occurred in this instance. It may be of course that the United States was acting on specific intelligence relating to one or more of those men it killed. Based on the publicly available evidence, it is more likely that a CIA error led to the deaths of five men whom it still refuses to acknowledge as civilians.

The implications of a US redefinition of 'combatant'

The reported re-interpretation by the CIA of the term 'combatant' as 'all military-aged males' in Waziristan and elsewhere has most likely enabled the United States to lower internal estimates of non-combatant casualties in its covert drone war. However, this process is not reflected either in external reporting or understanding. The majority of those killed at the 17 March 2011 *jirga*, for example, have been classified as civilians by the local population; by Pakistan's military, local, and federal government; and by local and international media. Subsequent field investigations and legal cases also continue to attest to the civilian non-combatant status of most of those killed. Additionally, discussions between this author and US, UK, and NATO military officials indicate that the CIA's reported re-definition of 'combatant' is unique. The British Ministry of Defence, for example, stated that:

> The UK has strict rules of engagement which are in accordance with both the Geneva and Hague conventions and are frequently updated in light of experience, intended to minimise the risk of casualties occurring.... When made aware of an incident or alleged incident, the UK follows a robust ISAF process to investigate. ISAF defines a civilian casualty incident as 'a civilian wounded or killed by ISAF'.[15]

Officials from the International Committee of the Red Cross (ICRC) have also indicated that any arbitrary redefinition of the term 'combatant' does not remove from belligerents their obligations under IHL and IHRL to minimise civilian casualties.[16] There are further implications. In unilaterally changing the definition of 'combatant', the United States may also have considered itself free to target individuals and groups in Pakistan who might normally have expected to be afforded the protections required under IHL or IHRL. This can be seen in the emergence of specific CIA tactics in Pakistan from 2009 onwards, some of which have been identified as possible war crimes by UN experts (Serle, 2012).

On 16 May 2009 Taliban militants had gathered in the Pakistan village of Khaisor, in anticipation of crossing the border into nearby Afghanistan to carry out insurgent activities. In the seventeenth reported drone strike of the Obama presidency, missiles initially destroyed a vehicle and a house. As militants and local villagers attempted to retrieve the dead and injured, drone operators – employing a novel tactic – struck again. As a local Taliban commander later reported, 'They were recovering bodies of the slain militants when the drone fired two more missiles, almost 12 minutes after the first attack. It was terrible as almost all the Taliban involved in rescue work were targeted' (Yusufzai, Khan, and Mujtaba, 2009). Four additional Taliban died: but so too did nine civilians from the village assisting in the rescue according to a field investigation by the BIJ for the *Sunday Times*, naming six men (Lamb *et al.*, 2012). From a US perspective the strike was doubtless viewed a success, employing for the first time a tactic designed to catch those militants drawn in to aid the casualties of an initial attack. As a Taliban commander noted at the time: 'We lost very trained and sincere friends. Some of them were very senior Taliban commanders and had taken part in successful actions in Afghanistan' (Yusufzai, Khan, and Mujtaba 2009). This first deliberate CIA attack on rescuers was not an anomaly. Between May 2009 and July 2011, at least 15 deliberate follow-up attacks on rescuers by the CIA were reported by credible news media, including the *New York Times*, CNN, Associated Press, ABC News, and Al Jazeera. The BIJ, in collaboration with the *Sunday Times*, conducted a four-month investigation into those claims. Field researchers found that on almost every occasion militants died. However in 10 of the reported attacks on rescuers the study also found credible, independently sourced, evidence of civilians killed – most of them named by researchers. The BIJ concluded that between 2009 and 2011 it was not possible for the CIA to target rescuers without a high risk of civilian fatalities (Woods, 2012a). In their report *Living Under Drones*, Stanford and NYU law schools also found extensive evidence of so-called 'double tap' strikes by the CIA, noting:

> Those interviewed for this report were acutely aware of reports of the practice of follow-up strikes, and explained that the secondary strikes have discouraged average civilians from coming to one another's rescue, and even inhibited the provision of emergency medical assistance from humanitarian workers.

Additionally the study noted: 'According to a health professional familiar with North Waziristan, one humanitarian organization had a "policy to not go immediately [to a reported drone strike] because of follow up strikes. There is a six hour mandatory delay" ' (Stanford and NYU, 2012: 76).

In August 2013 the BIJ reported on an additional short sequence of additional rescuer strikes in mid-2012, all of which appeared to have been focused on high-value targets. On this occasion the Bureau itself found no confirmation of civilian casualties, although independently both the legal charity Reprieve and Amnesty International identified and named eight civilian rescuers it said had died in one of the attacks (Amnesty International 2013). The BIJ concluded that a reduction in confirmed civilian casualties in 2012 'rescuer strikes' was likely due to a change in CIA targeting policy, coupled with wariness by civilians of participating in rescue operations (Woods, 2013). Although the United States has never formally responded to reports that it directly targets first responders, acting US ambassador to Pakistan Richard E. Hoagland told visiting US peace activists in 2012 that 'For at least the last several years that I have been here in Pakistan and more intimately associated with the knowledge of this [drone campaign], there was never any deliberate strikes against civilian rescuers' (Naiman, 2012).

A further and troubling CIA tactic was observed on 23 June 2009, when the Agency attacked a funeral near Makeen in South Waziristan. According to numerous press reports, as many as 5,000 people were present for the burial of mid-level Taliban commander Niaz Wali Mehsud, killed by a drone earlier that day. The CIA had hoped that Tehrik-e-Taliban Pakistan (Taliban Movement of Pakistan or TTP) leader Baitullah Mehsud would be present. He was not. Nevertheless the attack killed as many as 75 people, some eight to 50 of them reported to be civilians. (Yusufzai, Burki, and Mumtaz, 2009). Joby Warrick, a national security correspondent with the *Washington Post*, has alleged that Niaz Wali Mehsud was killed as 'bait' to lure Baitullah Mehsud to the funeral. In his book *The Triple Agent* – based on extensive briefings from US intelligence sources – Warrick claims that 'Agency officials in Khost and Langley watched on flat-screen TVs as cars arrived and mourners gathered', before ordering the attack (Warrick, 2011).

These tactics – the repeated and deliberate targeting of first responders, and the bombing of a funeral at which large numbers of civilians were present – marked a new chapter for the CIA's covert drone war in Pakistan from 2009 onwards. It is not known why (or how) such controversial tactics were approved. However, it seems likely that the adoption of the controversial definition of combatants as encompassing all military-aged males in a strike zone, as reported by the *New York Times*, was a contributory factor. Christof Heyns, the UN Special Rapporteur for extrajudicial executions, has expressed concerns regarding so-called 'double tap strikes'. Addressing the UN Human Rights Council Heyns noted: 'If civilian "rescuers" are indeed being intentionally targeted, there is no doubt about the law: those strikes are a war crime' (Serle, 2012). Heyns and fellow UN Rapporteur Ben Emmerson QC announced that they intended to examine the practice of 'double taps' as part of a broader investigation into armed drone use for the UN General Assembly (Emmerson, 2012).

So far this chapter has focused on the issue of how civilians are thought to be categorised by the United States, and to what extent this leads to a gulf between US claims of 'zero civilian casualties' and public sources claiming otherwise. However, there is also evidence that the CIA has deliberately withheld evidence of civilian deaths. On 22 April 2011, a CIA drone strike killed an estimated 25 people in a pre-dawn attack on a house and guest-house in Spinwan, North Waziristan. Of those killed, as many as nine were widely reported by international media – including the *New York Times*, the *Wall Street Journal*, and the BBC – to be women and children. In a highly unusual move, an anonymous US official denied to CNN at the time that any non-combatants had died in the attack, noting 'There is no evidence to support that claim [of civilian casualties] whatsoever' (Paton Walsh and Habib, 2011). A follow-up investigation by Associated Press a year later contradicted this. Based on extensive eyewitness reports the news agency noted:

> A funeral was held for the women and children, but the bodies of the militants were taken away, said [witness] Mamrez Gul. He said the women and children were relatives of the compound's owner, Gul Sharif, a militant commander loyal to Bahadur. He survived the attack, said two villagers, speaking on condition of anonymity.

AP also noted that 'A U.S. counterterrorism official in Washington said no women and children were observed in the compound before the strike' (Abbott, 2012). The Spinwan attack might have become just another incident in which credible public claims of civilian deaths were simply denied by the US. However, in April 2013 US news agency McClatchy published a major story based on leaked top secret US intelligence reports relating to the covert drone war. These were most likely briefing papers prepared for Congressional oversight committees,[17] and may therefore represent what the CIA wishes to be known about its targeted killing programme rather than a reflection of its internal knowledge. Whilst most of McClatchy's reporting focused on those militant groups targeted by the CIA, the news agency noted that the leaked papers also identified a single civilian killed within the Brennan time window – in the Spinwan strike. Citing a year-long period from September 2010 reporter Jonathan Landay noted: 'The reports estimated there was a single civilian casualty, an individual killed in an April 22, 2011, strike in North Waziristan, the main sanctuary for militant groups in Pakistan's tribal areas' (Landay, 2013). It is not clear why only one of the numerous non-combatants widely reported as having been killed by the CIA that day was listed as such by the Agency. What is certain, however, is that the CIA was explicitly aware of at least one civilian death having occurred in a drone attack, some nine weeks before Brennan's June 2011 speech claimed otherwise.

Conclusion

There are clear indications that the United States has sought to limit its IHL obligations by classifying almost all military-aged males killed in covert actions as 'combatants' in Waziristan and elsewhere, even when such a status cannot be supported on available evidence. This is not merely a statistical sleight of hand. In response to the leaked CIA intelligence briefing obtained by McClatchy, US National Security Council spokeswoman Caitlin Hayden issued a lengthy response noting the Administration's 'scrupulous adherence to the rule of law'. She insisted that the US pursues the distinction, 'the idea that only military objectives may be intentionally targeted and that civilians are protected from being intentionally targeted'. In addition Hayden noted that there must be a 'high degree of confidence … that innocent civilians will not be harmed' (Hayden, 2013). Hayden's comments, whilst welcome, also raise troubling questions. By the end of 2013 the United States had killed at least 2,800 people in drone targeted killings in Pakistan, Yemen, and Somalia. Some 440 or more of those killed are credibly reported to have been civilians.[18] How many of these covert drone strikes proceeded on the basis that no 'civilian non-combatants' were present, based on a unique interpretation of that term? How many of 400 or more civilians died because the United States chose to categorise them otherwise? And what implications might this have for an ongoing covert drone war which shows little sign of abating soon?

There is one additional point worth noting. The CIA's claim that it has access to a perfect weapon that no longer kills civilians is demonstrably absurd. Yet John Brennan's hyperbole obscured an important issue. Data from all casualty recording organisations has consistently shown that both numerically and proportionally, credible reports of civilian deaths in covert US drone strikes have fallen steeply since late 2010, in all theatres. The New America Foundation (NAF), for example, estimates that a minimum of 57 civilians died in Pakistan in 2011 (a 9 to 14 per cent non-combatant casualty rate). In 2012, NAF reports only five confirmed civilian deaths from a minimum of 222 killed – a 2 per cent rate. Data from the BIJ and Long War Journal indicates similar declining civilian casualty trends. There seems little doubt that both the Central Intelligence Agency and the Pentagon generally go to significant lengths to limit non-combatant casualties. Yet civilian deaths do still occur. A US airstrike on the Yemeni city of Rada'a in September 2012 killed 13 innocent people for example, mostly women and children. Only three months later would an anonymous US official grudgingly accept responsibility – and it remains unclear whether the US includes any of those killed at Rada'a among its list of civilian fatalities (Raghavan, 2012). This absence of transparency and accountability risks alienating those local communities the US would be better served winning over in its fight against Al-Qaeda and its allies, and risks undermining any long-term strategic aims of Washington's targeted killing project.

Notes

1 An early draft of this paper was presented at *Hitting the Target? How New Capabilities are Shaping Contemporary International Intervention,* University of Surrey, 12–13 July 2012.

2 Averaged data across three monitoring groups: Bureau of Investigative Journalism 2,505–3,584 estimated killed in 371 strikes (www.thebureauinvestigates.com/category/projects/drone-data); New America Foundation 2044–3377 killed in 360 strikes (http://natsec.newamerica.net/drones/pakistan/analysis); Long War Journal 2,687 killed in 343 strikes (www.longwarjournal.org/pakistan-strikes.php). All estimates accurate at 8 August 2013.

3 Even after Pakistan's government reported in 2013 that between 400 and 600 civilians had died in the campaign, an unnamed US official was insisting to ABC News: 'The notion that the United States has undertaken operations in Pakistan that resulted in the deaths of hundreds of innocent Pakistanis is ludicrous' (Ferran, 2013).

4 US covert strikes are tracked by a number of organisations, including the Bureau of Investigative Journalism, the New America Foundation, and the Long War Journal. As each employs a different methodology, there are variations in the numbers of strikes and casualties reported. This chapter employs BIJ data unless otherwise stated.
5 This figure is believed to include a small number of CIA drone strikes in Afghanistan during the 2001 invasion. Although initially attributed to an anonymous US counter terrorism official, the estimate was later sourced to the CIA itself by the *New York Times.*
6 The *LA Times* actually cites a strike on 22 August 2010. However this is US-datelined. The attack occurred 23 August according to Pakistan local time.
7 For example a senior US official told this author in an on-the-record though non-attributable comment by email on 17 July 2011: 'There haven't been any non-combatant casualties for about a year, and assertions to the contrary are wrong.'
8 Background comment to the author, July 2011.
9 Declaration of interest: Between April 2010 and June 2013 the author worked with the Bureau of Investigative Journalism as senior reporter, where he set up and led the BIJ's covert war investigation team.
10 The BIJ's drone casualty data is organic, reflecting the best present public understanding of each strike. As of August 2013, its data indicated that during the 'Brennan window', out of an estimated minimum 717 reported deaths between 65 and 102 civilians were now considered as having been killed, in 14 incidents. There were indications of an additional 0–90 civilian deaths in a further 15 attacks which could not be corroborated.
11 Details obtained via affidavits of survivors and relatives of those killed, and laid before the London High Court.
12 The Associated Press investigation was published across several linked articles on 26 February 2012. Background notes from AP, for example, can be found at www.ap.org/Content/Press-Release/2012/AP-analysis-of-casualties-from-US-drone-strikes-in-Pakistan-the-backstory.
13 Figures taken from monitoring organisation datasets, August 2013. New America Foundation presently undercounts the CIA's overall casualty estimate by some 400 deaths.
14 Unpublished interview conducted by the author, November 2013.
15 On the record email to author from Lt Col Nick Short, British military spokesman, 20 June 2012.
16 Non-attributable background conversation between this author and ICRC officials, 15 June 2012.
17 Details contained in the leaked papers in part match secret documents prepared for Pakistani officials and partially leaked in 2009 to the Pakistani newspaper *Dawn.* See 'Drone Strikes Killed High-Value Targets, US Tells Pakistan,' *Dawn,* 2 September 2009, at www.dawn.com/news/855031/drone-strikes-killed-high-value-targets-us-tells-pakistan.
18 Estimates from the BIJ to 31 December 2013.

References

Abbott, S. (2012) 'Report Sheds New Light on Drone War's Death Toll', *Associated Press* (26 February 2012), http://english.alarabiya.net/articles/2012/02/26/197138.html (accessed 12 January 2014).

Al-Aulaki, N. v, Obama, B., Gates, R., and Panetta, L. (2010) United States District Court for the District of Columbia (7 December), www.lawfareblog.com/

wp-content/uploads/2010/12/Al-Aulaqi-Decision-Granting-Motion-to-Dismiss-120710.pdf (accessed 15 October 2013).

Amnesty International (2013) *Will I Be Next? US drone Strikes in Pakistan*, www.amnesty-usa.org/sites/default/files/asa330132013en.pdf (accessed 12 January 2014).

Becker, J. and Shane, S. (2012) 'Secret "Kill List" Proves a Test of Obama's Principles and Will' (29 May 2012), www.nytimes.com/2012/05/29/world/obamas-leadership-in-war-on-al-qaeda.html (accessed 12 January 2014).

BIJ (Bureau of Investigative Journalism) (2011–present) Covert Drone War project, www.thebureauinvestigates.com/category/projects/drones (accessed 12 January 2014).

Brennan, J. (2011) 'Question and Answer Session Remarks Following an Address at Johns Hopkins University' (30 June). Relevant video clip and transcript archived at www.thebureauinvestigates.com/2011/07/19/in-video-obamas-drone-war (accessed 12 January 2014).

Brennan, J. (2012a) 'This Week', interview with George Stephanopoulos. Transcript, *ABC News* (29 April), http://abcnews.go.com/Politics/week-transcript-john-brennan/story?id=16228333&page=4#.T_MKl5GwV7c (accessed 12 January 2014).

Brennan, J. (2012b) 'Remarks of the Assistant to the President for Homeland Security and Counterterrorism as Prepared for Delivery', Woodrow Wilson International Center for Scholars, Washington, D.C., April 30. White House, www.lawfareblog.com/2012/04/brennanspeech (accessed 15 October 2013).

Capaccio, T. and Bliss, G. (2011) 'US Said to Reduce Civilian Deaths after Increasing CIA Pakistan Strikes', *Bloomberg News* (31 January 31), www.bloomberg.com/news/2011–01–31/u-s-said-to-reduce-civilian-deaths-after-increasing-cia-pakistan-strikes.html (accessed 12 January 2014).

Dawn Newspaper (2011a) 'Rare Condemnation by PM, Army Chief: 40 Killed in Drone Attack', *Dawn Newspaper* (18 March), http://dawn.com/2011/03/18/rare-condemnation-by-pm-army-chief-40-killed-in-drone-attack (accessed 12 January 2014).

Dawn Newspaper (2011b) 'Agencies: Zardari Urges US to Stop Drone Attacks', *Dawn Newspaper* (24 March), http://dawn.com/news/615531/us-should-stop-drone-attacks-says-zardari (accessed 12 January 2014).

Dilanian, K. (2011) 'CIA Drones May be Avoiding Pakistani Civilians', *Los Angeles Times* (22 February 2011), http://articles.latimes.com/2011/feb/22/world/la-fg-drone-strikes-20110222 (accessed 12 January 2014).

Editorial (2011) 'The CIA and Drone Strikes', *New York Times* (13 August 2011), www.nytimes.com/2011/08/14/opinion/sunday/the-cia-and-drone-strikes.html?_r=1 (accessed 12 January 2014).

Elliott, J., Currier, C., and Groeger, L. (2012) 'Interactive: How Obama Drone Death Claims Stack Up', *Pro Publica* (18 June 2012), www.propublica.org/special/how-obama-drone-death-claims-stack-up (accessed 12 January 2014).

Emmerson, B. (2012) 'Statement by the UN Rapporteur on Counter Terrorism and Human Rights Concerning the Launch of an Inquiry into the Civilian Impact, and Human Rights Implications of the Use of Drones and other Forms of Targeted Killing for the Purposes of Counter-Insurgency and Counter-Terrorism', archived by Foreign Policy, www.foreignpolicy.com/files/fp_uploaded_documents/130124_SRCTBenEmmersonQCStatement.pdf (accessed 12 January 2014).

Ferran, L. (2013) 'Kids Killed by Drones: U.S. Pushes Back', *ABC News* (24 July 2013), http://abcnews.go.com/Blotter/kids-killed-drones-us-pushes-back/story new?id=19759808 (accessed 15 October 2013).

Garlasco, M. (2012) 'NATO's Lost Lessons from Libya', *Washington Post* (12 June 2012), www.washingtonpost.com/opinions/natos-lost-lessons-from-libya/2012/06/11/gJQAhkAoVV_story.html (accessed 15 October 2013).

Hayden, C. (2013) 'Statement Cited in full in White House Responds to Drone Pressure: "We will continue to disclose as much as we can"', *McClatchy* www.mcclatchydc.com/2013/04/12/188419/white-house-responds-to-drone.html#storylink=cpy (accessed 12 January 2014).

Holder, E. (2012) 'Transcript: Attorney General Eric Holder Speaks at Northwestern University School of Law' (6 March 2012), archived at Public Intelligence, http://publicintelligence.net/attorney-general-eric-holder-speech-on-targeted-killing-of-u-s-citizens-full-transcript (accessed 15 October 2013).

Johnson, K. (2010) 'U.S. Defends Legality of Killing With Drones', *Wall Street Journal* (5 April 2010), http://online.wsj.com/news/articles/SB10001424052702303450704575159864237752180?mg=reno64-wsj&url=http%3A%2F%2Fonline.wsj.com%2Farticle%2FSB10001424052702303450704575159864237752180.html (accessed 12 January 2014).

Kayani, A. P. (2011) Press Release, Inter Services Public Relations, Pakistan Military (17 March 2011), www.ispr.gov.pk/front/main.asp?o=t-press_release&date=2011/3/17 (accessed 12 January 2014).

Lamb, C., Woods, C., and Yusufzai, R. (2012) 'Covert CIA Drones Kill Hundreds of Civilians', *Sunday Times* (5 February 2012), (paywalled) www.thesundaytimes.co.uk/sto/news/world_news/article867772.ece (accessed 12 January 2014).

Landay, J. (2013) 'Obama's Drone War Kills "Others", Not Just al Qaida Leaders', www.mcclatchydc.com/2013/04/09/188062/obamas-drone-war-kills-others.html (accessed 12 January 2014).

Long War Journal (2006–present) 'Charting the Data for US Airstrikes in Pakistan', www.longwarjournal.org/pakistan-strikes.php (accessed 12 January 2014).

Masood, S. and Shah, P. Z. (2011) 'CIA Drones Kill Civilians in Pakistan', *New York Times* (17 March 2011), www.nytimes.com/2011/03/18/world/asia/18pakistan.html (accessed 12 January 2014).

Naiman, R. (2012) 'Americans Press US Ambassador for End to Drone Strikes in Pakistan, and the Ambassador Responds', *Truth Out* (5 October 2012) (transcript and video), http://truth-out.org/news/item/11963-americans-press-us-ambassador-for-end-to-drone-strikes-in-pakistan-and-the-ambassador-responds (accessed 12 January 2014).

New America Foundation (2010–present) 'The National Security Studies Program', http://natsec.newamerica.net (accessed 12 January 2014).

New York Times (2011) Editorial: 'The CIA and drone strikes' (13 August), www.nytimes.com/2011/08/14/opinion/sunday/the-cia-and-drone-strikes.html?_r=1 (accessed 12 January 2014).

Obama, B. (2012a) 'Google Town Hall Debate' (31 January 2012), transcript and video archived at www.thebureauinvestigates.com/2012/02/01/analysis-obama-outs-secret-cia-drone-campaign-but-do-his-words-add-up (accessed 15 October 2013).

Obama, B. (2012b) 'Letter from President Obama to Congress under the War

Powers Resolution (1973)' (15 June 2012), www.whitehouse.gov/the-press-office/2012/06/15/presidential-letter-2012-war-powers-resolution-6-month-report (accessed 12 January 2014).

Obama, B. (2013) 'Remarks by the President at the National Defense University' (23 May 2013), The White House, www.whitehouse.gov/the-press-office/2013/05/23/remarks-president-national-defense-university (accessed 15 October 2013).

Pakistan Ministry of Foreign Affairs (2011) 'US Ambassador Summoned to Protest 17 March Drone attacks' (18 March 2011), www.mofa.gov.pk/romania/pr-details.php?prID=929 (accessed 12 January 2014).

Panetta, L. (2011) 'Transcript of Interview with Charlie Rose' *PBS* (6 September 2011), US Department of Defense, www.defense.gov/transcripts/transcript.aspx?transcriptid=4872 (accessed 15 October 2013).

Paton Walsh, N. and Habib, N. (2011) 'US Departs Pakistan Base', CNN (22 April 2011), http://edition.cnn.com/2011/WORLD/asiapcf/04/22/pakistan.drone.strike/?hpt=T2 (accessed 12 January 2014).

Press Release (2011) www.ispr.gov.pk/front/main.asp?o=t-press_release&date=2011/3/17 (accessed 12 January 2014).

Raghavan, S. (2012) 'When U.S. Drones Kill Civilians, Yemen's Government Tries to Conceal it', *Washington Post* (25 December 2012), www.washingtonpost.com/world/middle_east/when-us-drones-kill-civilians-yemens-government-tries-to-conceal-it/2012/12/24/bd4d7ac2–486d-11e2–8af9–9b50cb4605a7_print.html (accessed 12 January 2014).

Roggio, B. and Mayer, A. (2006–present) 'Charting the Data for US Airstrikes in Pakistan', *Long War Journal*, www.longwarjournal.org/pakistan-strikes.php (accessed 12 January 2014).

Serle, J. (2012) 'UN Expert Labels CIA Tactic Exposed by Bureau "A War Crime"', *Bureau of Investigative Journalism* (21 June 2012), www.thebureauinvestigates.com/2012/06/21/un-expert-labels-cia-tactic-exposed-by-bureau-a-war-crime (accessed 12 January 2014).

Shane, S. (2009) 'CIA to Expand Use of Drones in Pakistan', *New York Times* (3 December 2009), www.nytimes.com/2009/12/04/world/asia/04drones.html (accessed 12 January 2014).

Shane, S. (2011a) 'CIA is Disputed on Civilian Toll in Drone Strikes', *New York Times* (11 August 2011), www.nytimes.com/2011/08/12/world/asia/12drones.html?pagewanted=all (accessed 12 January 2014).

Shane, S. (2011b) 'Contrasting Reports of Drone Strikes', *New York Times* (11 August 2011), www.nytimes.com/2011/08/12/world/asia/12droneside.html?ref=asia (accessed 12 January 2014).

Stanford and NYU (2012) *Living Under Drones: Death, Injury and Trauma to Civilians from US Drone Practices in Pakistan*, Stanford Law School, International Human Rights and Conflict Resolution Clinic, NYU School of Law, Global Justice Clinic, www.livingunderdrones.org/wp-content/uploads/2013/10/Stanford-NYU-Living-Under-Drones.pdf (accessed 12 January 2014).

The News International (2011) 'NWA Tribesmen Protest Drone Attack Casualties', *The News International* (17 June 2011), www.thenews.com.pk/TodaysPrintDetail.aspx?ID=52979&Cat=7&dt=6/17/2011 (accessed 12 January 2014).

Warrick, J. (2011) *The Triple Agent: The Al-Qaeda Mole Who Infiltrated the CIA*, New York: Doubleday.

Warrick, J. and Finn, P. (2010) 'Amid Outrage over Civilian Deaths in Pakistan, CIA Turns to Smaller Missiles', *Washington Post* (26 April 2010), www.washingtonpost.com/wp-dyn/content/article/2010/04/25/AR2010042503114.html (accessed 12 January 2014).

Woods, C. (2011) 'US Claims of "No Civilian Deaths" are Untrue', *Bureau of Investigative Journalism* (18 July 2011), www.thebureauinvestigates.com/2011/07/18/washingtons-untrue-claims-no-civilian-deaths-in-pakistan-drone-strikes; additional dataset at www.thebureauinvestigates.com/2011/07/18/get-the-data-twenty-five-deadly-strikes (accessed 12 January 2014).

Woods, C. (2012a) 'Get the Data: Obama's Terror Drones', *Bureau of Investigative Journalism* (4 February 2012), www.thebureauinvestigates.com/2012/02/04/get-the-data-obamas-terror-drones (accessed 12 January 2014).

Woods, C. (2012b) 'Evidence in British Court Contradicts CIA Drone Claims', *Bureau of Investigative Journalism* (24 April 2012b), www.thebureauinvestigates.com/2012/04/24/british-legal-case-shines-fesh-light-on-civilian-drone-deaths (accessed 12 January 2014).

Woods, C. (2013) 'Investigation Finds Fresh Evidence of CIA Drone Strikes on Rescuers', *Salon* (1 August 2013), www.salon.com/2013/08/01/investigation_finds_evidence_of_cia_drone_strikes_on_rescuers (accessed 12 January 2014).

Woods, C. and Yusufzai, R. (2011) 'Get the Data: Twenty-Five Deadly Strikes', *Bureau of Investigative Journalism* (18 July 2011), www.thebureauinvestigates.com/2011/07/18/get-the-data-twenty-five-deadly-strikes (accessed 12 January 2014).

Wright, T. and Mehsud, R. (2011) 'Pakistan Slams US Drone Strike', *Wall Street Journal* (18 March 2011), (paywalled) http://online.wsj.com/article/SB10001424052748703818204576206873567985708.html (accessed 12 January 2014).

Yusufzai, M., Burki, I., and Mumtaz, M. (2009) 'No Prominent Militant Killed in Drone Attack', *The News International* (25 June 2009), www.thenews.com.pk/TodaysPrintDetail.aspx?ID=22926&Cat=13&dt=6/25/2009 (accessed 12 January 2014).

Yusufzai, M., Khan, M., and Mujtaba, H. (2009) '40 Militants Die in NWA Drone Strikes', *The News International* (17 May 2009), www.thenews.com.pk/TodaysPrintDetail.aspx?ID=22177&Cat=13&dt=5/17/2009 (accessed 15 October 2013).

6 Drones

Degrading moral thresholds for the use of force and the calculations of proportionality

Conway Waddington

Collateral damage is a focal point of much of the moral debate on drone strikes. This chapter argues that quantitative claims about collateral damage should not form the sole basis for that debate. Instead, attention is drawn to the qualitative principles guiding the decisions to pursue such strikes. The availability of drones may be subverting moral restraints that ought to govern decisions to employ force. Questionable adherence to norms of proportionality, and distinction between legitimate and illegitimate targets, are highlighted. The emergence of 'signature strike' and 'double tap' tactics are further symptoms of an ongoing moral deterioration.

This chapter seeks to provide clarity on some of the moral questions raised by drone strikes.[1] This investigation rests on the basis that much of the moral concern directed at the use of drones is related directly to their particular employment in the conduct of a military and political doctrine of targeted killing. The controversial practice of deliberately targeting individual militants forms an integral part of the United States' mix of counter-insurgency and counter-terrorist strategic approaches to fighting the 'war on terror'.[2] Through their prominent involvement in facilitating this doctrine, drones have become the focus of much of the critical argument against the doctrine to the point that the drones themselves often appear to be the central target for criticism.

Bearing in mind this context, this chapter attempts to evaluate two, overlapping, areas of morally relevant discussion: those issues related to the consequences of drone strikes, and those related to the reasoning processes behind drone strikes (and the targeted killing doctrine in general).[3] Moral concerns raised about collateral damage caused by drone strikes, the accountability and oversight provided for those strikes, and the moral legitimacy of a targeted killing doctrine, all fall within these two areas of discussion. It is worth noting that this chapter takes the position that many of these concerns are only a temporal aspect of drone strikes, contingent on the political qualities of the conflict in its current form. Instead, focus is directed toward the processes of ethical reasoning that lead to the establishment of a moral position on the use of drones for targeted killings, and the proportionality equations that govern those strikes. This chapter suggests that these decisions are affected by the availability of drones. However, the manner in which those drones are used is itself contingent on deeply seated norms within the US strategic security culture.

Debating the moral merits or failing points of drones or precision strike technologies in general is challenging because of disputes over what information is available about the harm caused through improper targeting or by collateral damage. The nature of the current conflicts where drone strikes are used generally precludes easily confirmable casualty figures, and the covert nature of their use has meant a lack of transparency on the part of their primary user, the United States. Yet it is precisely that aspect of collateral damage that is the most apparent moral impact of drone strikes, and is in turn one of the most significant aspects of the debate on the morality of the strikes themselves. This chapter explores the concept of collateral damage and pays particular attention to questions of proportionality with regards to calculations about acceptable levels of collateral damage, and the effect of drones on that calculation.

Without putting too much pressure on what is a very subtle distinction that is of questionable value outside of fields of philosophy, this chapter investigates the possible impact of the availability of drones on the *ethical* reasoning behind the *moral* judgements and choices made about drone strikes. Christopher Coker offers a neat summation of the relationship between morals and ethics: 'Moral thought is deeply concerned with the systematic examination of the relations of human being to each other. Ethics is concerned with how we should conduct relations with other moral agents. When codified by the state they become laws' (Coker, 2008: 7).[4]

In order to begin this investigation into the morality of drone strikes, it is first necessary to dispel some of the myths attached to drones, and also to explore the ways that broader discussion on targeted killing has been overshadowed by a fixation on drones. Following these steps, this chapter engages with the question of proportionality in relation to collateral damage claims, but show that quantitative claims about collateral damage should not form the basis for debate on drone strikes. Instead, attention is drawn to the qualitative principles guiding the decisions to pursue drone delivered targeted killing strikes, and illustrate how drone technology may be subverting the moral restraints that ostensibly ought to be applied to such decisions.

Drones are the focal point of criticisms of the targeted killing doctrine

To begin with, drones and associated precision strike weapons and systems represent enabling technologies. By removing a pilot from the direct controls of the vehicle, unmanned vehicles place the maximum distance between a dangerous environment and the vehicle's operator (Brooke-Holland, 2012: 5). Amongst the qualities that make precision strike doctrine highly favoured today, the aspect of reduced risk is especially prominent. It is largely the pursuit of this quality that has spurred development of unmanned platforms for surveillance and, later, for actual strikes. Keeping an aerial drone on station over a particular region until it is needed is highly useful. Such behaviour with manned aircraft is both risky, and limited by the physical requirements of a pilot or crew. It was and is still this capability of persistence which makes drones valuable as a surveillance system: their predominant but overshadowed primary occupation.

Armed drones utilising precision weaponry were used in conventional combat roles in 2001 in Afghanistan, but became a part of covert targeted killing on 3 November 2002 when a CIA-controlled Predator drone fired a missile to kill six suspected Al-Qaeda terrorists in Yemen (FAS, 2002).[5] The addition of deployable weapons to drones can be viewed as a natural progression, allowing quicker and more agile transition between reconnaissance, targeting, and execution. Their ability to loiter is central to the discriminatory power attributed to drones. Through persistent observation, drones are able to confirm a target and pick an opportune time to strike, in principle limiting collateral damage (Williams, 2013a). Where precision strike technology is an expression of a technological evolution toward providing an optimum application of decisive military force, drones represent a parallel evolutionary branch. As such, drone-delivered ordnance offers the benefits of precision guided weaponry, along with the political safety of not having personnel in immediate harm's way (Brennan, 2012).

The advantages of drones are limited to certain, contextually defined situations

Unmanned combat vehicles are the subject of many false claims and assumptions regarding their capabilities. To begin with, the popular use of the term 'drones' inaccurately promotes the idea that unmanned vehicles possess a greater degree of autonomy than they in actual fact currently possess. Increased autonomy is however a likely future development in drone technology and raises significant moral questions about turning war fighting, and thus killing, over to robotic control, but these matters are not germane to this current discussion. Furthermore, the vast majority of drones in use by the US at present are small sensor platforms, essentially remote-piloted basic airframes packed with various sensors. Even unmanned combat aerial vehicles (UCAVs) like the MQ-1 Predator or MQ-9 Reaper drones, which carry both sensors and weapons, are relatively simple and cheap airframes in comparison to strike fighters.[6] Even relatively advanced drones like the MQ-1 and MQ-9 are not stealthy, and are vulnerable to integrated air defence systems, interception by other aircraft, and jamming. As a result their use is constrained to permissive operating environments, where the airspace they seek to utilise is either not protected by air-defence systems, or the government that holds sway over that airspace has granted permission for drones to use it (Lewis, 2012: 299).

Drones are permitted to act over Yemen and the Pakistani Federally Administered Tribal Areas (FATA) at the invitation, and through the permission of those states, acknowledged or not (Crilly, 2010; Miller, 2013). Essentially, drones are able to function transnationally through US foreign relations, and not because they themselves are free to penetrate any airspace. Drones are nevertheless treated by media critics and, indeed, in some statements by Pakistani political actors, as sovereignty violating instruments. For its part, the UN continues to propagate this theme by its failure to fully investigate the, albeit complicated and often contradictory, relationship between the Pakistani leadership and the US regarding US drone strikes in the FATA.[7] Overestimating the capabilities of drones contributes to misunderstandings in the popular discourse on drones and drone strikes. This point is worth emphasising: the prominent role of drones in the execution of targeted killing strikes should not lead to a fixation on the drones themselves.

Bearing in mind this apparent tendency to focus on drones as the root cause of rights violations, prominent contributions to the debate on drone usage can also be accused of conflating concerns about the targeted killing programme, and perhaps the use of military force in general, with a specific critique of drones. The Stanford Law School and NYU School of Law publication 'Living under Drones' is one such well-cited criticism of drone strikes (Stanford/NYU, 2012) that serves as a central point of the contemporary drone debate. This report weaves an emotionally evocative narrative describing the impact of the drone campaign, with emphasis on the secondary effects of drones on the populace of heavily targeted areas. Substantial attention is directed toward such subsidiary effects as property damage, reduced access to education, and even the impact on burial traditions. In this latter case, specific reference is made to the devastating effects of Hellfire missiles on the corpses of victims of strikes which render traditional burial practices purposeless (Stanford/NYU, 2012: 55–73). A consistent feature of the 'Living under Drones' report is that these negative effects identified are not exclusive to drone-delivered strikes. However, it is these effects that are highlighted as part of a broader complaint made specifically against drones.

The 2013 Human Rights Watch report 'Between a Drone and al-Qaeda' similarly directs attention to drones where the concerns raised are not necessarily exclusive to, or even the result of drone strikes (HRW, 2013). Analysis of, for instance, a strike on a supposed terrorist training camp at Al-Majalah, Yemen, on 17 December 2009 notes the mistaken targeting of a nearby Bedouin village along with legitimate Al-Qaeda in the Arabian

Peninsula (AQAP) fighters. Perhaps most damningly a volley of Tomahawk cruise missiles was used for the attack. Such weapons are capable of significant precision but in the Al-Majalah strike, they were fitted with cluster bomb sub-munitions, a decidedly imprecise weapon (HRW, 2013: 66–72).

While such examples provide substantial evidence for the analytical conclusion that overzealous, possibly illegal force is being employed, the report persistently returns to the issue of drones. Some witness accounts are helpfully amended to include 'drones' where the witness has referred to 'planes' (HRW, 2013: 31–2). There is nothing dishonest in clarifying what appear to be simple misstatements, but it is nonetheless telling that such effort is made on what is essentially a side issue. The HRW report is ostensibly a critical analysis of the morality and legality of the US targeted killing campaign against AQAP in Yemen, yet it too appears to fixate, albeit subtly, on drones.[8]

Noting popular misunderstandings about drone capabilities, and the tendency to fixate on drones themselves instead of the broader question of the morality of the targeted killing doctrine does not mean, however, that there are not morally significant considerations about drone strikes.

Morally framed complaints against drone strikes are significantly tied to collateral damage

Arguments based around the morality of drone strikes tend to focus on consequences of such strikes. While the consequences of drone strikes are just one part of the philosophical spectrum of morality, they are nonetheless the most easily accessible. For instance, as a result of their technological refinement, precision weapon wielding drones are broadly considered to be capable of greater distinction between combatants and civilians than the shelling of a neighbourhood or carpet bombing. That capability serves as the basis for arguments for the positive morality of such systems. Indeed, Bradley Jay Strawser has made the argument that those states with precision capabilities, or access to drones, are morally compelled to use them in place of other, less discriminating systems (Strawser, 2010: 342–68). This is a persuasive argument that speaks to the moral asymmetry that is implied alongside technological asymmetry in contemporary asymmetric warfare. Importantly, Strawser's argument and other pro-drone arguments are based on the underlying assumption that drones do in fact cause less negative harm.

Anti-drone complaints predominantly hold that assumptions about drones being more capable of distinction between combatants and noncombatants are false. Indeed, moral criticism of drone strikes tends to revolve around the morally harmful consequences of the strikes. Simplified, the most readily accessible moral problem with targeted killing is exposed in situations when some of the people being killed are either not the intended targets, or are not themselves legitimate targets. The unintentional or accidental killing of innocents is an inescapable reality of the practice of war. Yet any decision to use lethal force must at least account for the possibility of such errors. Similar thinking forms the backbone of other arguments against, for instance, the death penalty, where, no matter how advanced the justice system, mistakes can still be made. In order to adhere to the moral expectations of contemporary armed conflict, efforts must then be taken to minimise collateral damage.

At this point, it is worth unpacking the concept of collateral damage. Neta Crawford suggests that collateral damage can be viewed in the context of three types of collateral damage incidents: first in this list are genuine accidents that were unforeseen and perhaps unforeseeable; a second category of systemic collateral damage where the harm is the unintended but plausibly foreseeable result of organisational choices and routines relating to rules of engagement and choice of weaponry; and a third category where harm to civilians is considered likely to be judged to be proportional according to the doctrine of double effect (Crawford, 2013). Taking these incidents into account, proponents of drone strikes would argue that due care is given to avoiding the foreseeable types of accidents, and mitigating the unforeseeable, precisely through the long-term and high-fidelity target observation made possible by drones. In principle, this argument appears sound and yet it remains hotly contested, notably because those capabilities offer only a potential means of limiting collateral damage, based on decisions over the manner in which drone strikes are employed. Discussion then turns to the actual results of drone strikes, and the major hurdle of attempting to anchor analysis of the consequences of strikes on hard empirical facts.

Quantitative analysis of collateral damage is hindered by unclear, obscured, or impossible to confirm reporting

A definitive empirical account of the realities of drone-strike collateral damage would do much to resolve debate over the relative effectiveness of drones, particularly in terms of their capability to mitigate collateral damage. Unfortunately, definitive empirical evidence for this is difficult, if not impossible to produce. This has not stopped critics and proponents alike from interpreting what information does arise. Major criticisms of the use of drones in the 'Living under Drones' report argue that drone strikes are responsible for significant harm to the civilian populace of the targeted areas, specifically referring to the Federally Administered Tribal Areas (FATA) of Pakistan, and to Yemen.[9] This particular report utilises civilian casualty figures collected and compiled by The Bureau of Investigative Journalism (BIJ).

The BIJ casualty figures set is one of the more damning perspectives on drone strikes, suggesting that for the period between June 2004 through mid-September 2012 (the period covered by the 'Living under Drones' report – Stanford/NYU, 2012) drone strikes killed between 474 and 881 civilians. This dataset suggests significant civilian casualties resulting from drone strikes. Amongst the various efforts to estimate and collate casualty figures, the lowest estimates of civilian death belong to the White House, which has maintained an official 'no-comment' stance, but has unofficially alluded to single-figure civilian casualty percentages (Becker and Shane, 2012).

Contrasted with the BIJ, the US Government (USG) has consistently argued that drone strikes are responsible for very few civilian casualties as a result of the technological benefits of drones and the efforts of those directing the drone strikes. Specific numbers are impossible to pin down as the USG refuses to release official drone strike figures. Such figures would, however, not by themselves be authoritative since comprehensive ground-level post-strike analyses are effectively impossible to conduct, for either the USG, or for independent observers.[10] Illustrative of the high regard in which the USG holds drones is however clear. On 29 June 2011, John Brennan, then the Assistant to the President for Homeland Security and Counter-terrorism, stated the following in relation to drone strikes:

> In fact I can say that the types of operations … that the US has been involved in, in the counter-terrorism realm, that nearly for the past year there hasn't been a single collateral death because of the exceptional proficiency, precision of the capabilities that we've been able to develop.
>
> (Williams, 2013: 97)

The absence of authoritative data has allowed the propagation of such casual portrayals as Brennan's. On the opposite side of the scale, Daniel Byman notes in an otherwise compelling argument against drone strikes, that for every militant killed, 10 civilians have died (Byman, 2009). While this is an opinion piece, which perhaps explains the casual manner of Byman's reference to those otherwise unsubstantiated figures, the substantial discrepancy between extremes presented between the likes of Byman and Brennan nonetheless speaks to the dangerous lack of objective verification of quantitative data on drone strikes.

Avery Plaw explored the specifics of the discrepancy in estimates for casualty figures resulting from drone strikes (Plaw, 2013). Plaw compared data from four prominent casualty datasets of drone strikes conducted against militants (consisting of members of groups such as Al-Qaeda, the Taliban, the Haqqani network, and others), in Pakistan (predominantly the North and South Waziristan regions). These datasets covered the period of 2004 to 2011. These sources included the Long War Journal, the University of Massachusetts' UMass DRONE project, the New America

Foundation, and the BIJ (Plaw, 2013). For that period, these four databases showed the percentage of civilians killed in strikes to be as follows:

The Long War Journal: 6.03 per cent
New America Foundation: 17.42 per cent
UMass DRONE: 3.86 per cent
The BIJ: 23.85 per cent.

The difference between the lowest and highest estimates shows a range of roughly one civilian killed for every 20 legitimate targets, or of one civilian killed for every four combatants.[11] Plaw notes that a major contributing factor to this discrepancy lies with the way that unknown casualties, those neither confirmed as civilians nor as terrorists, were counted. Under certain interpretations, particularly a 'guilt by association' line of thought unknown casualties may be counted as legitimate kills after the fact, thanks to the logic that those killed at the target sites containing terrorists, must have been terrorists themselves (Boyle, 2013: 7). Both the Long War Journal and the UMass DRONE datasets tended to identify unknowns as combatants and, hence, reflected non-combatant casualty ratios. Conversely, the other figures tended to count unknowns as non-combatants. Reporting biases amongst these datasets aside, the matter of distinguishing between combatants and non-combatants is a significant challenge to any effort to calculate the collateral damage effects of drone strikes.

Lack of empirical consensus opens the door to narrative interpretations

Brian Glynn Williams provides a useful overview of the confusion and misinformation that can arise in discussions of the effectiveness of drones in terms of discrimination, proportionality, and collateral damage based on questionable data (Williams, 2013a: 99–115).

Williams criticises anti-drone arguments, which he shows hinge on unsubstantiated empirical claims about their lack of efficacy in mitigating civilian casualties. In 2009 local media in Pakistan, which had previously displayed partisan tendencies, began to establish a narrative suggesting the drone strike campaign was in fact stunningly inaccurate (Williams, 2013a: 115). Because there is no specific evidence upon which they could be based, the implication is that these news reports must have been made up, and are therefore likely false. These media reports were in turn jumped upon by local politicians and leveraged at the national level as part of arguments against US violations of Pakistani sovereignty. Williams shows how these unsubstantiated claims were then cross-cited, gaining credibility by repetition (ibid.).

In particular, the Pakistani media narrative that drone strikes have been responsible for substantial civilian casualties gained traction through its recital in a May 2009 *New York Times* article by David Kilcullen and Andrew

Exum (Kilcullen and Exum, 2009). Both Kilcullen and Exum are respected national security figures, with Kilcullen having advised General David Petraeus. Kilcullen went on to repeat the Pakistan media figures in testimony to the US Congress, and in his book *Accidental Guerrilla* (2009). Kilcullen and Exum's analysis was in turn cited by Mary Ellen O'Connell in testimony to the US Congress as well as in several papers and articles on the topic, further entrenching this particular theme in the drone strike discourse (O'Connell, 2010).[12]

Ultimately, the empirical truth about the number of civilians being killed by drone strikes is likely to remain impossible to prove indisputably, regardless of increased oversight by the US. Even where there is consensus on the number of strikes and numbers of deaths, there remains a substantial fundamental dispute over the distinction between combatants and non-combatants. With these factors in mind, it is doubtful that attempts to justify or condemn drone strikes based purely on arguments based on quantitative claims about their capability to reduce civilian harm will ever achieve a definitive conclusion. Attention must therefore be directed at the qualitative aspects of drone strikes and the decisions to utilise the use of force and thereby undertake the risk of collateral damage.

Technological capability for collateral damage mitigation casts a spotlight on the morally uncomfortable question of proportionality and the doctrine of double effect

So far, the difficulties in turning to empirical evidence to settle debate over the merits of drones for mitigating collateral damage has been the focus of this chapter. It is therefore necessary to return to this debate by examining the underlying decision making processes that govern drone strikes. These processes are morally required to acknowledge the possibility of collateral damage, and to enact mitigating measure to limit that damage where possible. Where collateral damage is unavoidable, it is necessary to engage in the morally uncomfortable process of weighing up the proportionality of those moral harms with the good that will result from presumably disrupting some future terrorist attack. The implication voiced through anti-drone sentiment is that the availability of drone strikes in the tactical toolbox has affected the organisational ethos guiding the war on terror. Instinctively, this feels like a more solid basis for moral criticism, but the question is: how precisely do drones affect strategic decision making?

As discussed earlier, collateral damage exists on a sliding scale of morality. This sliding scale is based around the concept of proportionality, particularly in terms of the doctrine of double effect (DDE), which can be conceptually traced back to one of the progenitors of Just War Theory, Thomas Aquinas (Aquinas, 2006). This doctrine has played an integral role in Just War Theory, and is closely related to the moral criterion of proportionality, which is present in both the *jus ad bellum* and the *jus in bello*. In essence, the DDE sets out criteria under which the incidental negative consequences of an otherwise well-intentioned act will not delegitimise the entire act. Put differently, an act that is intended to produce certain morally good effects can still be considered morally justifiable even if additional negative consequences occur, providing certain criteria are satisfied.

These criteria stipulate that the act itself should be morally good or at least neutral, the negative consequences may not be a means to an end, or an end in and of themselves, and the good effects must outweigh the bad. In addition, due diligence must have been paid to attempting to mitigate those negative consequences. In the context of collateral damage, the causing of civilian deaths is a negative consequence that is only morally permissible where it abides by the criteria of the doctrine of double effect.

Thus, in the case of drone strikes, the strikes must be considered a morally good or at least neutral act, the collateral damage itself must not be intended (as a show of force for instance), and the killing of the target must be of sufficient importance that it outweighs the collateral harm caused. However, calculating the relative importance of the target against the harm of collateral damage is by no means a simple matter. As Michael Schmitt (2005: 457) adds:

> The principle of proportionality is often misapplied. For instance, in some cases the mere quantum of collateral damage and incidental injury causes critics to condemn a strike as disproportionate. However, the extent of harm and damage is relevant only in relation to the military advantage reasonably expected as the attack was launched.

Schmitt further notes that that standard is defined as a comparative concept of excessive force which is contrasted with merely extensive force, which is an absolute and therefore more readily accessible.

The tipping point implied here results in an assessment that 'excessive force' is being applied, but there remains a disconcerting lack of clarity on the specifics of what constitutes excessive force. Jason Wright suggests that because there is no conventional or customary definition of 'excessive', the decision must be made in accordance with reasonable military assessments and expectations, taking into account potential collateral damage (Wright, 2012: 839). Because of the confusion over the notion of excessive force and of proportionality in the *jus in bello*, it is little wonder then that so much effort is made to anchor pro- and anti-drone strike arguments on civilian casualty figures.

To add a further dimension to this equation, it should be considered that the surgical precision and persistence capabilities of drones serve to highlight the question of how proportionality is calculated. In principle, drones offer qualities and capabilities that reduce much of the guesswork behind strikes. However, while technological developments allow for the mitigation of unintentional collateral damage, drones can only reduce the risk of collateral damage, not eliminate it entirely. That risk of harm still exists, and must therefore be factored into the strike by strike decision to employ force (and indeed, in the broader decision to employ the targeted killing doctrine itself). Moreover, in some scenarios, collateral damage is not only foreseeable, it is expected, and requires the calculation of the relative moral worth of killing the target against the moral harm of killing civilians. This then shines a spotlight on the proportionality equation that is being employed to account for the double effect of the strike. Inevitably the question must be posed: how morally balanced is the weighting of civilian casualties in proportion to US national security? More specifically, it is necessary to consider the role of drones in conducting strikes as potential contributors to any biases or imbalances in that proportionality equation.

The targeted killing doctrine requires morally informed decision making

In a speech before the American Society of International Law, United States Department of State Legal Advisor Harold Koh described the US reasoning processes that occurred for approving a drone strike. The assessment is based on the imminence of the threat posed by the target, the sovereignty of the State involved, and the willingness and ability of that State to suppress the threat posed by the target. Moreover, Koh stated that the execution of a strike is governed by traditional Just War principles of distinction and proportionality (Koh, 2010). While this appears to be recognition by the State Department of the centrality of certain moral principles in executing drone strikes, it should be considered that the appropriation of Just War Theory-based language does not ensure against improper application of the underlying philosophical principles contained within.

In doubts about the authenticity of the moral standards being adhered to by the USG, Michael J. Boyle, a former security adviser for President Obama in the run-up to his 2008 election, expresses criticism of 'the Obama administration's growing reliance on drone strikes. This perceived reliance has adverse strategic effects that have not been properly weighed against the tactical gains associated with killing terrorists' (Boyle, 2013: 7). A further point to be made here is that the perceived tactical efficiency of drone strikes, if over-stated, can lead to pursuit of policies of targeted killing in settings where that efficiency is not strategically sound. Mistaking drone strikes for counter-terror panacea calls to mind the 'law of the instrument' commonly expressed as the 'golden hammer',[13] which, to paraphrase, suggests that if your favoured or only available tool is a hammer, it is possible for all problems to start to look like nails.

The same charge can be levelled against the US, in its pursuit of a military response and especially, drone strikes, as solutions to its security challenges. Where foreseeable collateral damage occurs, that damage is the result of deliberate decisions based on a perception of proportionality. In the case of drone strikes, that proportionality equation is subjectively tied to the security priorities and perhaps also the underlying psyche of US strategic security culture.

David Luban has noted the US war-model approach to counter-terrorism and the importance of this approach in characterising its overall foreign policy as part of that conflict (Luban, 2002: 9–14). The treatment of Al-Qaeda and the threat of terrorism as a military threat has been held as contentious since its inception. Yet a military response was inevitable, given the enormity of the strategic culture shock of the 9/11 attacks. Donald Rumsfeld offered a not atypical perspective in the aftermath of 9/11: "We did not start this war. So understand, responsibility for every single casualty in this war, whether they're innocent Afghans or innocent Americans, rests at the feet of the al Qaeda and the Taliban" (quoted in Wheeler, 2011). This sort of thinking could be a significant contributor to the proportionality equation employed by the US with regards to drone strike collateral damage.

Collateral damage resulting from strikes of questionable strategic value might be partially a result of the complacency arising out of political disengagement with the doctrine. Because drones are relatively low-risk in terms of potential political costs, it has previously been possible to turn them over to the covert world of plausible deniability and non-existent oversight. One significant aspect of this reduction in oversight is the allowance of reduced thresholds for use. The lack of risk to an operator further exploits this subversion of the traditional thresholds on the use of force. Where previously the decision to employ military force hinged at least partially on the risk of domestic fallout arising from casualties, drone warfare carries no such potential penalty (Cortright, 2012). Hence, a substantial moral restraint on the use of force is undermined by the opportunity presented by the reduced political accountability of drone strikes.

It does not however appear to make much sense to conclude that drones affect the proportionality calculations based merely on the fact that they are remotely operated. A state would be more likely to undervalue risk to civilians precisely where there is more threat to its personnel, not less as is the case with drones. A narrative has emerged to attempt to explain this phenomenon which suggests that drone operators, by virtue of their physical disconnection from the battlefield, are also morally disconnected from it (Alston, 2010: paragraph 84). However, accusations that drone operators are engaged in 'playstation warfare' appear naïve to the realities of modern warfare, which has certainly allowed for high-fidelity killing at distance long before drone strikes became prevalent (see any of the numerous gun camera videos from US attack helicopters eliminating insurgents or precision bombs striking tanks as evidence).

This discrepancy instead points to a vital theme which was alluded to in the early parts of this chapter: drones are an enabling tool. They enable the pursuit of a targeted killing campaign which, for tactical and political reasons is better conducted by unmanned vehicles which are able to loiter for long periods of time, and are politically low-cost in the event of a shoot down or crash. While drones themselves may not directly play a role in the devaluing of collateral damage in the proportionality equation, they do facilitate a strategic approach to national security which has taken on many morally questionable practices.

Signature strikes and double taps are symptoms of degraded moral perspectives

Drones have been implicated in two tactical employments that can be conceptually traced to the targeted killing rationale, but bear little resemblance in terms of moral justification. The first of these are signature strikes, which refer to an operational doctrine that emerged to guide drone strikes on targets of opportunity. Essentially, individuals located in regions considered to contain enemy combatants, exhibiting certain characteristics or showing certain behaviour, can lead to those individuals themselves being considered enemy combatants and legitimate targets. These targets do not need to be pre-identified through intelligence input, as with the 'personality strikes' that characterise targeted killing.[15] Such strikes raise significant moral questions.

Boyle (2013: 9) argues that '(T)he Obama administration's "guilt by association" approach implies an indifference to the combatant status of the potential victims that is at odds with the legal and moral responsibility to make this determination before killing them'. While drone capabilities for persistence and precision weapon delivery may aid such a tactic in minimising collateral damage to some extent, this tactic nevertheless sacrifices the pre-planning and intelligence-based rigour that is often summoned in defence of targeted killings. To employ such a tactic on the part of the US is to knowingly engage in actions that may result in collateral damage without any particular confidence that the intended targets are of particular value, let alone that they are legitimate targets (enemy combatants) at all.[16] Such a tactic borders on indiscriminate use of force, a clear moral violation.

The second phenomenon refers to instances where drones have directed follow-up missiles at recently struck target sites in order to strike at enemies attempting to remove fallen comrades from the sites (Woods and Lamb, 2012; Shane, 2012). One possible rationale behind such 'double-tap' strikes, as they are known, is that associates of the presumed targets are on site purely to remove evidence of their recently deceased compatriots having been there. According to this rationale, by removing bodies and other evidence, these individuals are contributing to the intelligence and propaganda wars that underlie counter-terrorism and counter-insurgency, and in so doing, become eligible targets themselves (Williams,

2013b). Such strikes are essentially based on guesswork over the arrival of legitimate targets amongst the civilians that might also gather at the strike site. Such guesswork calls into question the proportionality and discrimination of the follow-up strike.

Because there is no way of guaranteeing who precisely responds first to a drone strike, no follow-up strike can be proportionally calculated in terms of the doctrine of double effect, nor can it hope to uphold any reasonable standard of discrimination between civilians and combatants.

That such tactics have arisen is undoubtedly the result of the opportunity presented by drone technology. However, examination of the moral reasoning behind the employment of these tactics speaks to something that is happening at deeper organisational level. (It might be argued that the institutional thinking that leads to signature strikes is similar to the thinking that signed off on now discredited torture and rendition policies, but there is not space to pursue that hypothesis in this chapter.) The emphasis of this assessment falls on the moral judgements which inform the decision making processes of the targeted killing doctrine in terms of its scope, scale, continuance, and individual targeting policies. Drones facilitate these tactics, and indeed, may even have played a role in the germination of such moral perspectives, but to fixate on drone strikes alone is to miss the bigger picture.

Conclusion: morally contentious aspects of drones are a symptom of a deeper problem

Drones attract a disproportionate amount of attention in what should properly be a discussion on the moral appropriateness of the US's targeted killing doctrine. The consequences of drone strikes, particularly in terms of collateral damage, are at the forefront here. Unfortunately, empirical data on collateral damage has been, and is likely to remain difficult if not impossible to verify. This has not stopped pro- and anti-drone arguments that revolve around the efficiency or inefficiency of drones in reducing collateral damage. The result has been narrative-driven heavily biased analyses of the US drone strike campaigns in Pakistan and Yemen that do little to move the discourse forward. Arguably then, the discussion of the quantitative features of drone strikes can only carry the debate so far.

There is, however, room for investigation of the reasoning behind drone strikes in the context of collateral damage. It is the position of this chapter that the legitimacy of US drone strikes is morally controversial based on the questionable ethical reasoning that appears to be taking place behind evaluations of proportionality of an intended strike and the chance of collateral damage.

The availability of drones and their qualities of reduced operator risk and enhanced precision are important facets in that ethical reasoning.

However, these characteristics do not by themselves serve as vindication of the targeted killing doctrine. Paradoxically, it appears that the promise of superior discriminatory abilities of drones serves to highlight the uncomfortable realities of wilful decisions to incur collateral damage as a necessary part of achieving certain security goals. It is also reasonable to suggest that the reduced operator risk of drones plays at least some role in encouraging political complacency regarding oversight, and even in encouraging the use of force, where the domestic political pressure of potential casualties might otherwise dissuade it. Furthermore, a skewing of national security priorities toward the narrow, short term appears to be occurring. This may well be a strategic security issue, but the potential resultant lowering of thresholds for the use of force is also fundamentally a moral issue.

The availability and capabilities of drones appear to play a role in moral backtracking on previously established standards of restraint relating to proper discrimination between combatants and non-combatants. Where drones possess capabilities of persistent presence for careful observation to distinguish legitimate targets, an attitude of guilt by association appears to be present. Signature or double-tap strikes serve as particular evidence for the potential of a moral slippery slope in this regard. The suggestion that the enhanced personnel safety offered by drones and precision strike systems does not appear to account for this slippery slope, though, and the political ease of use of drones through reduced oversight may play a role. More tellingly, it appears that the calculation of proportionality between the positive morality of a particular strike against the moral harm of collateral damage is biased toward a unilateral focus on national security that may be flawed both in terms of overall effectiveness, but also in a more general sense of moral justice.

The question of whether the US counter-terrorism campaign serves to meet its intended outcome of providing for the safety and security of the US, has been an underlying, if not directly addressed subplot in this chapter. The key debateable point here is whether the killing of individuals deemed to represent imminent threats to the US effectively reduces the overall threat of terror attacks, and is therefore morally justifiable? If the legitimate targets being killed do not actually warrant a significant threat to the US, then any collateral damage is automatically disproportionate. Yet another question to consider is the discrepancy between the moral equations being calculated by the US, and by those within targeted countries. A bereaved family member in Yemen cannot be expected to accept the moral necessity of the death of one of their own as a collateral necessity for security the US from potential attack, not least of all when the basis for that argument rests on shaky ground.

Notes

1 The term drone is used here intentionally as a catch-all. Depending on nationality and service, all manner of acronyms pertaining to unmanned or remotely piloted, controlled, operated nature of variously armed or unarmed land, sea and air based vehicles, platforms and systems have been used. Common acronyms include Unmanned Aerial Vehicles (UAVs), Unmanned Ground Vehicles (UGVs), and Unmanned Combat Aerial Vehicles (UCAVs).

2 This terminology, introduced under the presidency of George W. Bush, has been dropped by the Obama administration. Nevertheless, the United States politically and militarily considers itself at war, as evidenced by the continued reliance on the 2001 Authorisation of the Use of Military Force (AUMF). Under the law, the US government reserves the right to employ war powers, as self-defence, against a broadly defined threat of terrorism, embodied by Al-Qaeda and its affiliates.

3 The specific codes of ethical conduct here refer to bodies of moral thought that together comprise the *jus in bello* and *jus ad bellum*. These are moral frameworks that have been codified in the bodies of international law, and in the laws of armed conflict found in the Geneva Conventions.

4 Illustrative of this, the notion of legitimacy exists as both a moral concept and as a legal one. In the legal world, legitimacy is positioned alongside the notion of legality. The actions of a state might be described as legal, but illegitimate and *vice versa*. Crucially, democratic legitimacy is heavily dependent on the moral quality of the decisions being made (Peter, 2010). Legitimacy is predicated on the consent of the governed, effectively determined by putative norms, a morality-based concept. Throughout this chapter, I refer to legitimacy in the moral sense.

5 Incidentally, the first drone strike in Pakistan occurred 18 June 2004. The Bureau of Investigative Journalism. (2011) 'The Bush Years 2004–2009'.

6 This cheapness is not without contention either. The relative low cost of larger drones like Reaper or Predator is challenged by the relatively high maintenance costs and presumably shorter lifespan. That said, the loss of a drone to enemy fire is significantly cheaper than the loss of a combat aircraft, particularly if the potential loss of a pilot is factored in.

7 Joshua Foust, a fellow at the American Security Project criticised UN Special Rapporteur Ben Emmerson's investigation into US drone strikes in Pakistan on the grounds that Emmerson failed to properly investigate the relationship between the Pakistan military and intelligence service, and the US. Foust, J. (2013) 'The Islamabad drone dance', Need to know – On PBS.

8 It is however worth noting that perception counts. Retired US General Stanley McChrystal, the former International Security Assistance Force (ISAF) commander, cautioned repeatedly that drones are hated on a 'visceral' level, and that they are seen to embody the perceived arrogance of the United States: Alexander, D. (2013) 'Retired general cautions against overuse of "hated" drones', Reuters.

9 Moreover, 'Living under Drones' notes that drones have resulted in harm to the civilian populace in other ways than purely through missile strikes. Psychological and sociological harms caused to the civilian populace of drone-targeted regions, where local inhabitants now fear to gather in large groups and many show symptoms of post-traumatic stress disorder (PTSD).

10 The BIJ admits that it is only able to tally strikes and casualty figures based on imperfect information that comes out of the FATA. The 'Living under Drones' report, along with other similarly 'anti-drone' reports (such as the HRW and Amnesty International reports), all acknowledge that the information coming

out of Pakistan is imperfect, that independent verification is difficult, and that often, media reports and second-hand information must be utilised – this is the reality of any combat-zone reportage, but it could certainly be argued that that level of uncertainty is especially high in attempting to catalogue drone strike figures in the FATA.

11 There is no specific ratio prescribed in any legally binding document regarding acceptable behaviour in terms of combatant to non-combatant deaths. Indeed efforts to derive historical casualty ratios have been unable to produce conclusive results. There is some consensus that one of the more comprehensive estimates for historical trends suggests a 50 per cent ratio (Eckhardt, 1989: 97).

12 By way of a counter-point Christine Fair has conducted research that directly disputes the statistics cited by Kilcullen and Exum (Fair, 2010).

13 This cognitive bias is attributed mostly, but not exclusively, to psychologist Abraham Maslow.

14 The killing of specifically targeted individuals based on prior intelligence work are sometimes referred to as 'personality strikes', particularly in contrast to signature strikes (Stanford/NYU, 2012: 12–13).

15 Micah Zenko (2013b) notes that a May 2013 speech by President Barack Obama at the National Defence University, which introduced sweeping changes to drone strike policy in terms of increasing transparency and oversight, failed to mention anything about the future of signature strikes.

References

Alston, P. (2010) 'Report of the Special Rapporteur on Extrajudicial, Summary or Arbitrary Executions, Philip Alston', UN General Assembly Human Rights Council.

Aquinas, T. (2006) *Summa Theologica, Part 1 (Prima Pars)*, from the Complete American Edition, Project Gutenburg, www.gutenberg.org/ebooks/17611 (accessed 7 January 2014).

Becker, J. and Shane, S. (2012) 'Secret "Kill List" Proves a Test of Obama's Principles and Will', *New York Times*, 29 May 2012.

Boyle, M. (2013) 'The Costs and Consequences of Drone Warfare', *International Affairs*, 89: 1.

Brennan, J. (2012) 'The Efficacy and Ethics of U.S. Counterterrorism Strategy', Remarks made on 30 April 2012 at the Woodrow Wilson International Center for Scholars, www.wilsoncenter.org/event/the-efficacy-and-ethics-us-counterterrorism-strategy (accessed 7 January 2014).

Brooke-Holland, L. (2012) *Unmanned Aerial Vehicles (Drones): An Introduction*, Library of the House of Commons.

Byman, D. (2009) 'Do Targeted Killings Work?', Brookings Institute, www.brookings.edu/research/opinions/2009/07/14-targeted-killings-byman (accessed 7 January 2014).

Coker, C. (2008) *Ethics and War in the 21st Century*, London: Routledge.

Cole, C. (2012) 'The Drone Wars Briefing', Drone Wars UK, dronewars.net/2012/01/01/the-2012-drone-wars-briefing (accessed 7 January 2014).

Cortright, D. (2012) 'License to Kill', Cato Unbound, www.cato-unbound.org/2012/01/09/david-cortright/license-kill (accessed 7 January 2014).

Crawford, N. (2013) *Accountability for Killing: Moral Responsibility for Collateral Damage in America's Post-9/11 Wars*, New York: Oxford University Press.

Crilly, R. (2010) 'Wikileaks: Pakistan Privately Approved Drone Strikes', *The Tele graph*, www.telegraph.co.uk/news/worldnews/wikileaks/8172922/Wikileaks-Pakistan-privately-approved-drone-strikes.html (accessed 7 January 2014).

Eckhardt, W. (1989) 'Civilian Deaths in Wartime', *Security Dialogue*, January 1989, Sage Publication, 20, 1: 89–98, http://sdi.sagepub.com/content/20/1/89.extract (accessed 7 January 2014).

Fair, C.C. (2010) 'Drone Wars', *Foreign Policy*, www.foreignpolicy.com/articles/2010/05/28/drone_wars#sthash.FdLxS8NR.dpbs (accessed 7 January 2014).

Federation of American Scientists (FAS) (2002) 'RQ-1 Predator MAE UAV Overview', Intelligence Resource Programme, www.fas.org/irp/program/collect/predator.htm (accessed 7 January 2014).

HRW (Human Rights Watch) (2013) 'Between a Drone and Al-Qaeda', www.hrw.org/reports/2013/10/22/between-drone-and-al-qaeda (accessed 7 January 2014).

Kilcullen, D. (2009) *The Accidental Guerrilla*, London: Hurst and Company.

Kilcullen, D. and Exum, A. (2009) 'Death from Above, Outrage down Below', *New York Times*, www.nytimes.com/2009/05/17/opinion/17exum.html?pagewanted=all (accessed 7 January 2014).

Koh, H. (2010) 'The Obama Administration and International Law', Speech at the Annual Meeting of the American Society of International Law, Washington, D.C., www.state.gov/s/l/releases/remarks/139119.htm (accessed 5 January 2014).

Lewis, M. (2012) 'Drones and the Boundaries of the Battlefield', *Texas International Law Journal*, 47, 2: 294–314.

Luban, D. (2002) 'The War on Terrorism and the End of Human Rights', *Philosophy and Public Policy Quarterly*, 22, 3: 9–14.

Melzer, N. (2008) *Targeted Killing in International Law*, New York: Oxford University Press.

Miller, G. (2013) 'In Interview, Yemeni President Acknowledges Approving U.S. Drone Strikes', *Washington Post*, www.washingtonpost.com/world/national-security/yemeni-president-acknowledges-approving-us-drone-strikes/2012/09/29/09bec2ae-0a56-11e2-afff-d6c7f20a83bf_story.html (accessed 7 January 2014).

O'Connell, M. (2010) 'Unlawful Killing with Combat Drones: A Case Study of Pakistan, 2004–2009', Notre Dame Law School Legal Studies Research Paper 9, 43: 1.

Peter, F. (2010) 'Political Legitimacy', *Stanford Encyclopaedia of Philosophy*, http://plato.stanford.edu/entries/legitimacy (accessed 7 January 2014).

Plaw, A. (2013) 'Counting the Dead: The Proportionality of Predation in Pakistan', in Strawser B.J. (ed.) *Killing by Remote Control: The Ethics of an Unmanned Military*, New York: Oxford University Press.

Schmitt, M. (2005) 'Precision Attack and International Humanitarian Law', *International Review of the Red Cross*, 87, 859: 455–66.

Shane, S. (2012) 'US Said to Target Rescuers at Drone Strike Sites', *New York Times*, www.nytimes.com/2012/02/06/world/asia/us-drone-strikes-are-said-to-target-rescuers.html?_r=0 (accessed 7 January 2014).

Stanford/NYU (2012) 'Living Under Drones', www.livingunderdrones.org/wp-content/uploads/2013/10/Stanford-NYU-Living-Under-Drones.pdf (accessed 7 January 2014).

Strawser, B. (2010) 'Moral Predators: The Duty to Employ Uninhabited Aerial Vehicles', *Journal of Military Ethics*, 9, 4: 342–68.

Strawser, B. (ed.) (2013) *Killing by Remote Control: The Ethics of an Unmanned Military*, New York: Oxford University Press.

Walzer, M. (2013) 'Targeted Killing and Drone Warfare', *Dissent – A Quarterly of Politics and Culture*, www.dissentmagazine.org/online_articles/targeted-killing-and-drone-warfare (accessed 12 January 2014).

Wheeler, N.J. (2011) 'Protecting Afghan Civilians from the Hell of War', *Social Science Research Council Essay Forum*, http://essays.ssrc.org/sept11/essays/wheeler.htm (accessed 7 January 2014).

Williams, B. (2013a) *Predators: The CIA's Drone War on al Qaeda*, Washington D.C.: Potomac Books.

Williams, B. (2013b) 'New Light on CIA "Double Tap" Drone Strikes on Taliban "First Responders" in Pakistan's Tribal Areas', *Perspectives on Terrorism*, 7: 3.

Woods, C. (2011) 'US Claims of 'No Civilian Deaths' Are Untrue', Drone strike in Pakistan – The Bureau of Investigative Journalism www.thebureauinvestigates.com/2011/07/18/washingtons-untrue-claims-no-civilian-deaths-in-pakistan-drone-strikes (accessed 7 January 2014).

Woods, C. and Lamb, C. (2012) 'Obama Terror Drones: CIA Tactics in Pakistan Include Targeting Rescuers and Funerals', The Bureau of Investigative Journalism, www.thebureauinvestigates.com/2012/02/04/obama-terror-drones-cia-tactics-in-pakistan-include-targeting-rescuers-and-funerals (accessed 7 January 2014).

Wright, J. (2012) '"Excessive" Ambiguity: Analysing and Refining the Proportionality Standard', *International Review of the Red Cross*, 94: 886.

Zenko, M. (2013a) *Reforming U.S. Drone Strike Policies*, Council on Foreign Relations, Special Report No. 65.

Zenko, M. (2013b) 'How does the Recent Shift in U.S. Drone Policy Impact "Signature Strikes"?', Council on Foreign Relations, 11 June 2013.

7 Terrorising terrorists – the *jus ad bellum* of drone operations in Pakistan

Tobias Ruettershoff

Introduction

The chapter analyses the *jus ad bellum* of U.S. drone operations in Pakistan. The U.S. administration has recently admitted to conducting such operations, defending them as a legal and legitimate response to terrorists. It is argued that the decision to apply the laws of armed conflict to these operations is an important judgement call that the U.S has to make in each individual case. Yet, if requirements of necessity, immediacy, and proportionality are met, drone operations would fall under Article 51 of the UN Charter, which has developed in recent years to allow self-defence against armed attacks by non-state actors.

Up to and including the present day, the modern nation-state is perceived as the referent object of all International Relations and thus the entity of International Law is modelled around it. However, the emergence of violent non-state actors (i.e. 'terrorists') and their actions against other states, e.g. in the terrorist attacks of 11 September 2001, challenge these foundations, as there are seemingly no clear legal norms about how to deal with them. Within the last decade, benefitting from technological advance, Unmanned Aerial Vehicles (UAVs), often known as 'drones', have increasingly become the United States' (U.S.) preferred military tool for fighting Al-Qaeda and other terrorist groups.[1] First used by the U.S. as an offensive weapon during the invasion of Afghanistan in November 2001, they are now used in various military conflicts and crisis regions around the world, such as Iraq, Libya, Somalia, and Yemen, yet mostly in the Afghanistan–Pakistan region (BIJ, 2013).

Drones are appealing, because they pose minimal to no threats to U.S. military personnel, but their use has sparked a highly controversial debate about their use to kill selected individuals suspected of terrorism or affiliation with terrorist groups (Murphy, 2008; Ramsden, 2011; Schmitt, 2012; Shah, 2010; Sterio, 2012). Yet the legal issues pertaining to UAVs are by no means unique. For example, the debate about whether an incursion into Pakistan to strike extremist fighters operating from there amounts to a breach of the country's sovereignty is not just limited to drones, as the same issues would apply to manned fighter jets or Special Forces operating across borders. Furthermore, the disagreement between legal scholars as to whether such operations should be considered as international armed conflict, non-international armed conflict, or rather as a law-enforcement activity is an important question as it determines the body of law to be applied, yet employment of UAVs as a specific tool does not affect this determination. Indeed, even the complex legal and ethical issues raised by 'targeted killing' operations, such as the legal status of the target or the circumstances of the killing, would have to be discussed under the same legal precepts if they were conducted by car bombs or sniper attacks (Schmitt, 2012: 596f.).

Still, global attention has focused on UAVs and so does this chapter. After outlining the U.S. justification for the strikes, I consider whether the laws of war can be applied to the transnational fight against terrorism with drones or if the latter has to be seen under human rights law, which would put much stronger restraints on the use of force. I argue that the law-enforcement paradigm is very difficult in dealing with terrorist threats originating from areas which are no longer under host government control. Therefore, it is a reasonable option – which the US has chosen – to assess the drone strikes under the laws of armed conflict and not a law-enforcement regime, as the latter does not allow the necessary means to be used. This raises important questions about the violation of Pakistan's sovereignty, as Article 2 (4) of the United Nations (U.N.) Charter entails a general prohibition of the use of force. An exception to this prohibition is Article 51. In the traditional interpretation of the International Court of Justice (ICJ) and customary international law, Article 51 does not warrant a right to self-defence against non-state actors, unless they are imputable to another state. Yet, as is argued here, in recent years there has been a change. Article 51 is now more and more interpreted in a way that states must be able to defend themselves against terrorist threats, hence allowing transnational use of force against terrorists. However, to limit the risk of abuse of Article 2 (4), the strict rules of customary law need to be applied. These are the principles of immediacy, necessity, and proportionality.

In this chapter, I am not attempting to defend the cross-border use of drones by the U.S. against terrorists based in Pakistan. In fact, I maintain that, against the background of the principles of customary law, the overwhelming majority of U.S. incursions into Pakistani territory are illegal under international law because they do not meet these strict requirements. However, drone attacks are not per se illegal, if the U.S. can prove – in each individual case – that an operation meets the requirements, as there must be the possibility for a state to defend itself against an ongoing or impending attack, even from non-state actors. This leads to two considerations for further research. On the one hand, the problematic legal nature of U.S. drone strikes points to a problem with the current law as there is too much risk of abuse. On the other, if a small state (Pakistan) consents to drone operations by a stronger state (U.S.), because it is aware that the stronger state would violate its territory anyway, this raises the question of whether the territory of smaller states becomes an open battlefield for militarily stronger states.

The position of the United States

Although the drone operations in Pakistan were initiated under President George W. Bush, he only oversaw 48 drone strikes in his second presidency. Barack Obama, who "vowed to fight terrorism with greater effectiveness and moral rectitude than his predecessor" (McCrisken, 2013: 97), had by August 2013 ordered another 311 strikes (New America Foundation 2013). Although U.S. and international media had widely reported on drone attacks, President Obama only justified his stance on the policy on 30 January 2012, defending the attacks against alleged terrorists as 'precise, precision strikes against al-Qaeda and their affiliates', arguing that 'this is a targeted, focused effort at people who are on a list of active terrorists who are trying to go in and harm Americans, hit American facilities, American bases' (Obama, 2012). Despite this apparent new openness of the Administration to talk about drone strikes, especially in Yemen and Somalia, the CIA drone programme in Pakistan is still classified. Several government officials have also argued in favour of the attacks, making constant reference to the Administration's viewpoint that these actions were in accordance with domestic and international law. John O. Brennan then Obama's chief counter-terrorism advisor and now director of the CIA, stated that

> the principle that guides all our actions, foreign and domestic – we will uphold the core values that define us as Americans, and that includes adhering to the rule of law … that includes covert actions, which we undertake under the authorities provided to us by Congress … all our actions – even when conducted out of public view – remain consistent with our laws and values.
>
> (Brennan, 2011)

In fact Harold Koh, then legal adviser to the U.S. State Department, was the first official to admit to targeted killing practice, which he did in a speech at the American Society of International Law Annual Meeting on 25 March 2010: '[I]t is the considered view of this Administration [...] that U.S. targeting practices, including lethal operations conducted with the use of unmanned aerial vehicles, comply with all applicable law, including the laws of war' (Koh, 2010). He defended the killings by offering several reasons supporting their legality, citing both domestic (Authorization for Use of Military Force Against Terrorists, AUMF) and international law to prove that the U.S. is engaged in an armed conflict with terrorist groups such as Al-Qaeda and the Taliban and that the attacks are in accordance with the laws of war. First, for Koh, targeted operations are legitimate because the terrorists are members of an enemy group at war with the U.S. Second, drones are an appropriate tool for the killings as long as their use is consistent with the laws of war. Third, the attacks, although directed at individuals, do not violate the U.S. ban on assassinations[2] because the state using lawful weapons in self-defence or in an armed conflict is not illegal and hence does not constitute assassination. Fourth, and most importantly, in its target selection process the U.S. government adheres to the principles of 'distinction' and 'proportionality'. Distinction 'requires that attacks be limited to military objectives and that civilians or civilian objects shall not be the object of the attack'. Proportionality, in turn, 'prohibits attacks that may be expected to cause incidental loss of civilian life, injury to civilians, damage to civilian objects, or a combination thereof, that would be excessive in relation to the concrete and direct military advantage anticipated' (Koh, 2010).

Despite the attempts to justify the strikes, they are criticised by many. Philip Alston, U.N. Special Rapporteur on Extrajudicial, Summary, or Arbitrary Executions, has said that targeted killings 'violate straightforward applicable legal rules' (U.N., 2010: 3). In the following sections, the legal challenges of targeted killings and the U.S. drone programme in Pakistan are more closely analysed. I contend that targeted killings are not per se illegal. If performed under certain circumstances and strict rules of engagement, they could potentially comply with the *jus ad bellum*[3] criteria, if they are part of an armed conflict. Hence, I first assess whether the transnational use of drones against non-state actors can be seen as a shooting war or rather a law-enforcement issue and then look at the requirements of the law of armed conflict.

War paradigm or law-enforcement paradigm?

The U.S. maintains that targeted killings are a legitimate tool in the 'war on terror', whose legality must be assessed in terms of the laws of war; critics refer to them as 'extra-judicial' and would prefer to see these actions assessed under a domestic law-enforcement model, which rests mainly on international human rights law (IHRL) (Kretzmer, 2005: 174). Human rights activists, non-governmental organisations (NGOs) such as the International Commission of Jurists (2004), or U.N. institutions such as the Economic and Social Council (2004), dismiss the claim that the laws of armed conflict can be applied to terrorism, because they believe that the real threat is an over-reaction by adopting measures that run against human rights. They either assume that the existing constraints of international human rights law on anti-terror measures do not prevent an effective counter-terrorism strategy, or if they do, the dangers to human rights resulting from relaxation of these constraints are far greater. Thus, their categorisation of targeted killings as 'extra-judicial' implies that the legality of such strikes has to be judged by a strict law-enforcement model which rests on human rights law (Kretzmer, 2005: 175f.).

Under this law-enforcement regime, the intentional use of deadly force by state agencies can be justified only in very limited conditions, primarily self-defence or defence of others, or the execution of a death sentence after a fair trial: 'A killing is only legal to prevent a concrete and imminent threat to life, and, additionally, if there is no other non-lethal means of preventing that threat to life' (Wuschka, 2011: 898). The state must respect and ensure the right to life and due process of law.[4] Under a law-enforcement paradigm, utilizing human rights law 'it is never permissible for killing to be the sole objective of an operation' (U.N., 2010: para. 33). Hence, U.S. drone strikes would be illegal.

The U.S. argues that targeted killings have to be seen under a war paradigm, where under international humanitarian law the use of lethal force is more permissible than under the law-enforcement paradigm. The central argument is that transnational terrorists have military abilities and means (and a willingness to use them) which can only be countered with military measures. Indeed, the actions of individual terrorists and terrorist organisations and the violence they inflict go far beyond that of ordinary criminals. Nearly 3,000 people died in the terrorist attacks of 11 September 2001. Israel lost over 900 citizens in the first five years of the Second Intifada. In their attacks, 'the terrorist organisations use military means par excellence, whereas the common denominator of them all is their lethalness and cruelty. Among those means are shooting attacks, suicide bombings, mortar fire, rocket fire, car bombs, et cetera' (Supreme Court of Israel, 2006: para. 16).

Moreover, it is difficult to take the law-enforcement model or environment as the basis for the fight against terrorists, as there is often an 'unusual, and at times insurmountable obstacle to indicting them' (Meisels, 2004: 304). The ability to build a strong case against a suspected terrorist, proving him or her guilty, is inherently difficult. Evidence is often hard to get, particularly in states which either refuse to cooperate in law-enforcement or are simply incapable of it. Suspects in custody are an additional problem for the states. Juries and witnesses may be intimidated by the terrorists, making the trial even more problematic (Finn, 1991: 99). The detention of terrorists may also lead to further attacks or hostage-takings to enforce their release (Machon, 2006: 40). Admittedly, the fact that it is difficult to build a case against a suspect does not justify a targeted killing. If terrorists were operating from within the U.S., it would be impossible to argue for a targeted drone attack, given U.S. national and local law-enforcement capabilities (Heymann and Kayyem, 2004: 68). This is different for terrorists operating from abroad as it renders one of the principles of the law-enforcement model invalid: 'that the suspected perpetrator is within the jurisdiction of the law-enforcement authorities in the victim state, so that an arrest can be effected' (Kretzmer, 2005: 179).[5]

In the case of transnational terrorism, the victim state is reliant on the assistance of the other state where the terrorist is located. If this state is willing and able in terms of its governing capacity to arrest the suspect, then IHRL would again disallow targeted killings (Anderson, 2009: 30). However, if a state is unwilling or unable to handle terrorist threats originating from its soil, the situation is different. Setting aside this issue of unwilling states – as it would raise legal issues other than those pertaining to a discussion on international armed conflict – the question is what options are available in a state that is willing to cooperate and arrest a suspected terrorist, yet does not have sufficient governing capacity to effectuate this. Inaction is obviously not an option as those people would remain in the areas that law-enforcement agencies cannot reach (Ramsden, 2011: 401).

One option would be a large-scale operation. This may require significant ground forces and it is extremely doubtful whether a sovereign state such as Pakistan or Yemen would cooperate with the U.S. Even in a 'failed state' (such as Somalia), where a central government is non-existent on a large scale, such actions can present insurmountable obstacles, as seen in the U.S. raid to apprehend accomplices of Somali warlord Mohammed Farrah Aidid in Mogadishu, on 3 October 1993, resulting in the death of 18 U.S. soldiers and about 500–1,000 Somalis (Bowden, 1999). Such tremendous collateral damage and loss of life could have possibly been avoided by a more kinetic approach (Machon, 2006: 39f.).[6] Looking at Pakistan, a key ally of the U.S. in the fight against terrorism, it is clear that the state is generally willing (albeit hesitantly) to assist the U.S. in chasing and arresting suspected terrorists. However, the government of Pakistan generally does not have control over the FATA, from where most of the Taliban and Al-Qaeda elements operate (Murphy, 2008: 10). Prior to 9/11, the largely autonomous provinces were almost entirely ruled by local Pashtun tribal groups. With the U.S. invasion of Afghanistan, Pakistan's government was urged to regain control over the areas, which resulted in some reluctant counter-terrorism operations by the Pakistani army. They did not really eliminate terrorist strongholds but caused significant civilian casualties and local resentment (Jones, 2008: 59f.).

Hence, it is obvious that the law-enforcement paradigm has major difficulties in dealing with terrorist threats originating from areas which are no longer or have never been under government control. On the one hand, law-enforcement agencies of the victim state are often unable to operate in foreign countries. On the other hand, if the host government attempts to use law-enforcement measures, the backlash from the militants is more likely to create more of a quagmire than if a concerted and limited military action, such as a drone attack, were taken. As I have shown, there is no axiomatic answer to the question whether drone strikes against terrorists have to be assessed under the laws of armed conflict and not a law enforcement regime. In effect, this is an important judgement call that the U.S. (or any other state) has to make and be prepared to explain and justify in each individual case. In the following section, I look at the issues that are raised by applying the laws of armed conflict as the assumed basis for the legal evaluation of drone strikes in Pakistan, in particular about the violation of Pakistan's sovereignty.

Jus ad bellum

Self-defence against non-state actors

The question is whether international law allows the resort to force against terrorists in a foreign country. Since 1945, the U.N. Charter has become the dominant international set of rules for the resolution of armed conflict. A 'cornerstone' is the prohibition of the use of force in Article 2 (4) (ICJ, 2005: para. 148). It is a peremptory norm (*ius cogens*) and although states have often violated it, their reference to exceptions to this prohibition to justify the use of force shows their legal conviction about its existence (Gazzini, 2005: 124f.). 'Force', within the meaning of Article 2 (4), does not comprise any form of coercive power but in the prevailing view only armed force (Dinstein, 2005: 86). Moreover, this force has to be used in the 'international relations' of the states, which means that only cross-border use of force would be seen as an offence (Randelzhofer, 2002: para. 29). Although the scope of 'force' is restricted to armed force, the prohibition is extensive. Any use of military force on foreign territory or against a foreign state, no matter how minor, constitutes a violation. Article 2 (4) not only covers large-scale military operations but also small-scale force with military means (Schmitz-Elvenich, 2007: 39), including drone attacks. Hence, targeted killing operations by UAVs generally qualify as a breach of the prohibition of the use of force according to Article 2 (4).

However, such a breach does not occur if there is expressive consent by the state whose territorial integrity is violated (Schachter, 1984: 1644f.). Every day, states consent to actions by other states which would be a violation of international obligations without such consent, such as overflight rights for airspace, but also the authorisation to conduct a public investigation in a foreign state. Any rising frequency of terrorist attacks increases the probability that states will give their consent to wide-ranging counter-terrorism measures by a foreign state on their soil, especially if that state is itself incapable of acting. The consent by the state can indeed be reluctant, based on the insight that the intervening state would act anyway (Schmitz-Elvenich, 2007: 34). This could be a serious concern, depending on how much and with what means the intervening state pressures the host state. In the case of cross-border operations into Pakistan, the U.S. legal justification relies heavily on Pakistani consent (Entous *et al.*, 2012), yet it is contested how 'willing' Pakistan actually is. Publicly, the Pakistanis denounce the strikes and, on several occasions (Dawn, 2011; Pakistan Ministry of Foreign Affairs, 2013), have made diplomatic protests.

Yet, there seems to be a silent cooperation between the intelligence services, including selection of targets (Ignatius, 2008), and even an implicit acknowledgement on the political level; according to leaked cables (Wikileaks, 2011), Pakistani Prime Minister Gilani said to US Ambassador Anne W. Patterson: 'I don't care if they do it as long as they get the right people. We'll protest in the National Assembly and then ignore it.' As the *Washington Post* has noted, Pakistan 'formally protests such actions as a violation of its sovereignty, the Pakistani government has generally looked the other way when the CIA has conducted Predator missions or U.S. troops respond to cross-border attacks by the Taliban' (Whitlock, 2008). Indeed, Pakistan's reluctance to condemn the drone attacks, at least in the past, is a strong indicator of tacit consent (Murphy, 2008: 19).[7]

Still, even if such consent may be assumed, basing the legality of drone operations on it is troublesome. Consent, whether explicit or implicit, is not a legally binding commitment. Moreover, Pakistan's reluctance to act may not be based in the belief that it is secretly accepting the drone strikes, but simply because it has no (military) capabilities to stop the U.S. from conducting such strikes. This is indicated by the fact that the Pakistani government has taken a stronger, more offensive stance towards the strikes in recent years, which makes even an implicit acquiescence, as described above, more unlikely. Also, even if there is free consent, it could be limited by certain requirements, such as prior notification or a limited scope of operations. Thus, the legality would be confined to the requirements set by Pakistan and any other action would infringe on Art. 2 (4) (Murphy, 2008: 21).

But the U.N. Charter provides two exceptions to the general prohibition of the use of force under Article 2 (4). First, under Chapter VII, the Security Council (SC) may authorise any measure necessary to 'maintain or restore international peace and security'. This possibility is not explored further here. Although it is not necessarily irrelevant for the issue of drones, as the U.S. could use the argument of 'threats to international peace and security' if the self-defence argument was not available, a SC resolution would still constitute clear legal justification for such strikes. Second, Article 51 stipulates that

> Nothing in the present Charter shall impair the inherent right of individual or collective self-defence if an armed attack occurs against a Member of the United Nations, until the Security Council has taken the measures necessary to maintain international peace and security.

Thus, Article 51 is a justification which does not cure the violation of Article 2 (4) but removes its illegality (Schmitz-Elvenich, 2007: 50).

Whilst Article 2 (4) clearly prohibits the use of force by states against other states, Article 51 is less clear about a state's self-defence against an armed attack from non-state actors. Based on a traditional reading, one might argue that it was written on the basis of rights and obligations between states and that any action of self-defence has to be in response to an armed attack imputable to another state (Myjer and White, 2002: 7). The question whether armed attacks by non-state actors grant a right to self-defence is not new (see, for example, Bowett, 1972; Brownlie, 1958; Sofaer, 1986) and it has become the focus of a more intense discussion in recent years. There have been several instances where the ICJ has had to give its understanding of self-defence and the interpretation of Article 51. In the past it has displayed a rather restrictive understanding. In the *Military and Paramilitary Activities* case (also known as *Nicaragua v. U.S.*; ICJ, 1986), the Court stated that armed attacks by non-state actors could principally allow self-defence measures, but that a non-state actor would have to be under 'effective control' of another state (ICJ, 1986: paras 109, 115). Yet, the mere 'provision of weapons or logistical or other support' was not a sufficient requirement for an armed attack (ICJ 1986, para. 195). At that time, this strict view was in line with that of the international community on cross-border actions against non-state actors, as seen in the condemnation of Israel for incursions into Lebanon (Levenfeld, 1982) and South Africa for attacking insurgents in the Angolan Civil War (Alexandrov, 1996: 180). In its *Advisory Opinion on the Israeli Wall* (ICJ, 2004), the Court rejected Israel's claim of self-defence against terrorists based on, first, the fact that Israel had not claimed that these attacks were imputable to another state and, second, that the attacks were not transnational, as they had occurred within Israeli occupied territory (ICJ, 2004: para. 139). Hence, based on the Court's interpretation of the *jus ad bellum*, Article 51 does not warrant a right to self-defence against non-state actors, unless they are imputable to a state.

However, the ICJ decision was not undisputed at the time of the judgement (see, for example, dissenting opinions of Judges Oda, Schwebel, and Jennings, ICJ, 1986; Sofaer, 1986) and especially since the early 2000s, the traditional and restrictive interpretation of Article 51 by the ICJ has come under considerable criticism (see, for example, Cassese, 2001; Krajewski, 2002). In terms of actual practice in the last decades, many states have deviated from their former strict stance and embraced a broader reading of Article 51, exercising or asserting a right of self-defence against armed attacks by non-state actors which are not imputable to another states. Cases include Iran (U.N., 1999), Russia (Wandscher, 2006: 194), Colombia, (OAS 2008), Turkey (Rubin and Tavernisse, 2008), the U.S. (Lobel ,1999), and again Israel (Zimmermann 2007). Especially the 9/11 terrorist attacks seem to have influenced the international community's view on self-defence against non-state actors (Krajewski, 2002; U.N., 2001a). As Cassese noted:

> in a matter of a few days, practically all states (all members of the Security Council plus members of NATO other than those sitting on

> the Security Council, plus all states that have not objected to resort to Art. 51) have come to assimilate a terrorist attack by a terrorist organization to an armed aggression by a State, entitling the victim state to resort to individual self-defence.
>
> (Cassese, 2001: 996f.)

Not just the international community, but also the ICJ seems to have deviated from its former strict stance on the issue. In the 2005 case *Armed Activities on the Territory of the Congo* (ICJ, 2005), instead of repeating the position of the Wall Case advisory opinion, the Court opined that there is 'no need to respond to the contentions of the Parties as to whether and under what conditions contemporary international law provides for a right of self-defence against large-scale attacks by irregular forces' (ICJ, 2005: para. 147). In separate opinions, Judges Kooijmans (2005: para. 28) and Simma (2005: para. 11) stated that the Court's reading of Article 51 as restricted to self-defence only against other states was outdated, given the SC decisions and state practice.[8] Although systematic state practice is one of the requirements of customary law, it is not sufficient to show that Article 51 allows self-defence against armed attacks from non-state actors (Murphy, 2008: 35). Nevertheless, as practice in the last couple of decades has shown,[9] states do respond militarily to such attacks and they are being less admonished for it by the international community than in the early 1980s.

In this sense, I argue that it is not sufficient to justify exceptions to Article 2 (4) on non-written justifications (Stahn, 2004: 842), but there should be a possibility for the victim state to defend itself against terrorists operating from foreign territory, if the host state is not willing or able to take the measures itself and if the operation is solely confined to effective counter-terrorism operations and against terrorist facilities. A flexible interpretation of Article 51 (Schmitz-Elvenich, 2007: 121) would allow for this within the existing legal framework. Hence, based on the development of state practice in the last couple of decades and stronger criticism of the ICJ my argument is that Article 51 has to be interpreted differently to include force against non-state actors. However, as there is an obvious chance that any possible form of justification might be abused to use force and undermine Article 2 (4), each operation has to be analysed under the strict constraints of customary international law.

Constraints of customary international law

As then U.S. Secretary of State Daniel Webster made clear in 1842, the exercise of self-defence as an inherent and autonomous right required the demonstration of the 'necessity of self-defence, instant, overwhelming, leaving no choice of means and no moment for deliberation' (Webster, 1842). The customary principles of immediacy, necessity, and proportionality as reflected in the *Webster doctrine* have to be analysed in this context.

First, *immediacy* is concerned with temporal limits to self-defence. If the attack is already taking place, it is clear that responses need not be limited to 'on-the-spot-reaction[s]' (Dinstein, 2005: 220) but also be allowed a certain time after the attack has taken place. This also applies to terrorist actions, if they appear as a series of attacks. If there is a consistent pattern to them, they can be seen as an *accumulation of events* which counts as one armed attack (Cassese, 1989: 596). A different situation occurs if an armed attack is anticipated and self-defence measures are taken before the actual attack takes place. In terms of drone attacks, this would mean that they could be conducted before terrorists have attacked. There are two different categories of anticipatory self-defence: pre-emptive and preventive.[10] *Pre-emptive* use of force means that a state reacts to an imminent threat. A state that pre-empts basically has the choice between hitting first or getting the first blow. In essence, pre-emption mandates certainty about an armed attack in the very near future (Gray, 2007: 13). In scholarly discourse, the prevailing opinion is that a state can react to such an imminent threat. When 'danger [is] immediate, and, as it were, at the point of happening' (cited by Kegley and Raymond, 2003: 389), pre-emptive action is allowed and legitimate. Hence, there seems to be an *opinio juris communis* that such a right exists (Krajewski, 2005: 12).

In contrast, a *preventive* use of force entails the use of military force to remove a potential future threat which might appear on the horizon but is not yet imminent (Kegley and Raymond, 2003: 388). The state often even has the choice to tolerate such a possible incident (Gray, 2007: 13). It is widely acknowledged by scholars that this mode of military action is unlawful and inadmissible as it would result in a 'bottomless legal pit' (Kaplan and Katzenbach, 1961: 213).[11] Concerning U.S. drone attacks in Pakistan, this would require that there is a clear and present danger of imminent attack against the U.S. or its troops. If this is not the case and a drone attack was conducted preventively, e.g. to kill a Taliban leader who has not engaged in planning or executing an attack, this would make it illegal, as the principle of immediacy is not fulfilled. But if the U.S. can convincingly prove that a drone strike has stopped an ongoing or imminent attack by non-state actors, they would meet that principle.

Second, the principle of *necessity* requires that the state only responds to an armed attack with the use of force when other defensive measures are not available or not sufficient (Dinstein, 2005: 210). For example, in its *Oil Platforms case*, the ICJ (2003: para. 74f.) found that the U.S. had neither complained to Iran about the military activities allegedly prepared on these platforms nor shown that they were a threat, so that the necessity requirement had not been fulfilled. For U.S. cross-border operations into Pakistan, this would raise the question of whether law-enforcement measures or diplomatic measures to eliminate the terrorist threat had been attempted (Dinstein, 2005: 237). However, as demonstrated above, this option is often not feasible, because terrorists operate militarily and the states they operate from are mostly unwilling or incapable of cooperating in law-enforcement. Moreover, the time for a surprise attack is often too limited to include the host state and the transmission of secret information to others might endanger the mission.[12] Still, the intervening state has the burden of proof to show that softer means were not available or sufficient (Schmitz-Elvenich, 2007: 157).

Third, within the scope of *proportionality*, the permissibility of the use of force has to be seen in terms of extent and intensity. On the one hand, the self-defence measure has to be necessary to remove the armed attack, but on the other hand it has to be in an appropriate relation to the current or future threat (Schmitz-Elvenich, 2007: 158). The defending state must not use more force than needed to reach the goal of the measure, i.e. the removal of the terrorist threat (Schachter, 1989: 222f.). Taking the Oil Platforms case as an example again, the ICJ noted that an appropriate response to a (proven) missile attack by Iran on a U.S.-flagged ship could include the destruction of an Oil Platform in the Gulf, as long as that platform had been involved in assisting the attacks. Hence, the Court opined that proportionate self-defence was not limited to infliction of a missile attack in response, nor limited to the targeting of missile site itself. Yet, the Court also noted that in an isolated attack such as the single mining of a ship (damaging but not sinking it) the destruction of numerous vessels and aircraft in return, as well as oil platforms, is clearly disproportionate (ICJ, 2003: para. 75). Considering transnational terrorism, a proportionate response would thus be not only to attack the particular terror cell responsible, but also the superior terrorist network to prevent further attacks (Schmitt, 2002: 7). With various attacks, the proportionality of a countermeasure might be doubted, as it could exceed the dimension of each attack. However, if there is a consistent pattern to the attacks and they are seen as an *accumulation of events*, a single self-defence measure is still proportionate, even if it is more damaging (e.g. in casualties) than each single attack (Schmitz-Elvenich, 2007: 160).

Assessing the necessity and proportionality of U.S. cross-border operations against terrorist targets in Pakistan, it is important to focus on the two different kinds of attacks that seem to have prompted those operations: the attacks of 9/11 and the more recent terrorist operations from Pakistan into Afghanistan (Murphy, 2008: 39). First, the attacks of 9/11 gave the U.S. a right of self-defence against the threat posed by Al-Qaeda (as granted in U.N. SC Resolutions 1373 and 1386). Against the background that the U.S. had asked for the extradition of Bin Laden for years and considering that the Taliban regime in Afghanistan was actively harbouring and supporting Al-Qaeda activities in Afghanistan (Dorronsoro, 2008: 459f.), the measures of the U.S. seemed immediate, necessary, and proportionate.[13] However, most experts would disagree that this is also valid for cross-border operations into Pakistan. Not only is Pakistan a third country that was not harbouring Al-Qaeda terrorists at the time of 9/11, but it has actively cooperated with the U.S. in finding and arresting or killing terrorists in the aftermath. Although the U.S. might have a certain distrust for Pakistan's counter-terrorism policy (Miller *et al.*, 2013) this does not grant the unilateral use of force into Pakistan. Moreover, especially after the death of Bin Laden, Al-Qaeda in Pakistan is now considerably weaker, making them operationally far less threatening than they were in 2001 (Murphy, 2008: 40f.).[14]

Second, cross-border raids by militants from Pakistan into Afghanistan to attack U.S. and other ISAF forces are more difficult to assess under the *jus ad bellum* requirements of necessity, immediacy, and proportionality. In terms of necessity and immediacy, the U.S would have to pursue all available channels to obtain Pakistani cooperation in handling the issue. If the authorities in Pakistan were unwilling or unable to stop such attacks from their territory, a self-defence measure could be deemed necessary. Such a measure would further be proportionate, if they directly and immediately targeted the militants initiating the original attack, using a comparable scale of force as compared to attacking other militants in other places using greater amounts of force (yet, as stated above, this could be proportionate if designed to remove the overall threat of cross-border attacks on coalition forces in Afghanistan). Hence, *jus ad bellum* regulations do not generally preclude U.S. drone strikes in Pakistan in response to militant attacks. Yet, the constraints of customary international law condition the manner in which the U.S. operations are conducted, potentially leading to considerable differences in the scope and intensity of self-defence.

Conclusion

As I have shown, state practice in the last couple of decades, accompanied by stronger criticism of former ICJ rulings regarding the right to use force, such as UAV attacks, against non-state actors not attributable to a foreign state, has, in fact, caused Article 51 of the U.N. Charter to develop to include force against non-state actors. Whilst it is difficult to justify exceptions through other non-written justifications, a flexible, modern interpretation of Article 51 that includes non-state actors would generally allow cross-border drone operations within the existing framework of international law. But this whole setting points to a larger problem. A flexible interpretation of Article 51 could be an easy argument to circumvent the prohibition of the use of force in Article 2 (4). As I argued, this could be prevented if each drone operation is analysed under the constraints of customary international law. However, this cannot be a permanent solution. Current international law is still based on a traditional, territoriality-based, understanding of sovereignty and is particularly problematic in connection with contemporary 'new conflicts' between a state and non-state actors on foreign territory. Although this is still the basic axiom of international law and new conflicts have to be measured against it, a reformulation of Article 51 to clearly include non-state actors and the caveats of customary international law would make the assessment of drone attacks much easier.

Regarding the current U.S. drone policy, as this analysis has proved, most of the strikes would have to be considered illegal, because they do not meet the strict caveats of international law. With approximately 400–900 civilians killed since 2004 (BIJ, 2013), the necessity and proportionality of the overwhelming majority of drone attacks are in serious doubt. Moreover, the requirement of immediacy is not always visible, as the U.S. sometimes extends the spatial and temporal reprisals against terrorists. Yet, under close scrutiny, some of the attacks might be considered an appropriate form of self-defence against cross-border terrorist attacks, if the principles of immediacy, necessity, and proportionality were observed.

The capabilities provided by precision strike technologies such as drones have dramatically increased since the beginning of the millennium. Their use in conflicts has attracted much debate and criticism. This chapter has attempted to shed some light on the legal aspects of the use of drones against non-state actors in cross-border operations, affecting the sovereignty of another nation. As I have shown, drone strikes can be legally justified if terrorist actors plan and/or conduct a military-style attack against another nation from the soil of a host nation which is unwilling or unable to prevent this. Yet, this raises another question for future research. If Article 51 allows the use of force in self-defence against non-state actors (whether in the proposed flexible interpretation, or as an amended regulation) by a stronger state on the territory of a weaker state, based on the latter's inability or unwillingness to fight terrorism determined by the stronger state, this could lead to a situation where the territory of smaller states is an open battlefield for militarily stronger states. This problem is beyond the scope of this chapter, which has focused on determining the legality or illegality of cross-border use of drones in conflicts with terrorists, exemplified by the U.S. use of drones in Pakistan.

Notes

1 The complex relationships between Al-Qaeda, and the Afghan and Pakistani Taliban are beyond the scope of this chapter.

2 By this he meant Executive Order 12333, signed by President Reagan in 1981, which states that 'No person employed by or acting on behalf of the United States Government shall engage in, or conspire to engage in, assassination' (White House, 1981: 2.11).

3 I acknowledge that there is a separate set of arguments about their compliance with *jus in bello*, but these are beyond the scope of this chapter.

4 See also Art. 6 (1) of the International Covenant on Civil and Political Rights (ICCPR, U.N. 1966); Art. 4 of the American Convention on Human Rights (OAS, 1969); Art. 4 of the African Charter of Human and People's Rights (OAU 1981); and Art. 2 (1) of the European Convention for the Protection of Human Rights and Fundamental Freedoms (Council of Europe, 1950).

5 As an *argument a contrario*, Alston in his report on extrajudicial killings states that 'less-than-lethal measures are especially appropriate when a State has control over the area' (U.N., 2010: para. 77).
6 Although conducted by the military, the aim of *Operation Gothic Serpent*, which resulted in 'The Battle of Mogadishu', was the capture (and not killing) of two of Aidid's lieutenants (Bowden, 1999).
7 As a matter of fact, the U.S. had stationed part of its drone fleet at Shamsi Air Base in Baluchistan, until it was ordered to leave after US-led NATO forces accidentally attacked and killed 24 Pakistani soldiers on the Afghanistan-Pakistan border (BBC, 2011).
8 Kooijmans had already noted in a separate opinion on the *Wall Case* that the 9/11 bombings made a new assessment of Article 51 necessary (see Separate Opinion of Judge Kooijmans, 2004: para. 35).
9 See examples above.
10 As noted by several authors (e.g. Krajewski, 2005: 4; O'Connell, 2002: 2) there is a certain inconsistency in terminology. Some consider pre-emptive use of force as a reaction to threats which are not imminent, analogous to the US National Security Strategy of 2002 (White House, 2002). Others term this use of force 'prevention'. I am using the latter interpretation, since it seems to be the most stringent (so do, for example, Gray, 2007; Kegley and Raymond, 2003).
11 As UN Secretary-General Kofi Annan (U.N. 2002) said:

> Any state, if attacked, retains the inherent right of self-defense under Article 51 of the Charter. But beyond that, when States decide to use force to deal with broader threats to international peace and security, there is no substitute for the unique legitimacy provided by the United Nations.

12 As was the case in the killing of Osama Bin Laden, where the U.S. did not inform Pakistan about the upcoming raid out of fear that he might be warned (Walsh, 2011).
13 And they remain so. Whilst the invasion itself was an international armed conflict between the U.S. and Afghanistan, the U.S. and its allies in the NATO-led International Security Assistance Force (ISAF) are now engaged in a non-international armed conflict mandated by U.N. SC Resolution 1386 (U.N., 2001).
14 However, Al-Qaeda has become stronger in other regions of the world, especially Africa (Zimmermann, 2013).

References

Alexandrov, S.A. (1996) *Self-Defense against the Use of Force in International Law*, The Hague: Kluwer Law.

Anderson, K. (2009) *Targeted Killing in Counterterrorism and Law*, Working Paper of the Series on Counterterrorism and American Statutory Law, a Joint Project of the Brookings Institution, the Georgetown University Law Center, and the Hoover Institution. Washington, D.C.: Brookings Institution, http://papers.ssrn.com/sol3/papers.cfm?abstract_id=1415070 (accessed 28 August 2013).

BBC (2011) 'Pakistan Orders Nato and US Review After Deadly Border Strike', BBC (27 November 2011), London, www.bbc.co.uk/news/mobile/world-asia-15905777 (accessed 28 August 2013).

BIJ (2013) *Covert Drone War*, London: Bureau of Investigative Journalism.

Bowden, M. (1999) *Black Hawk Down: A Story of Modern War*, New York: Atlantic Monthly Press.

Bowett, D.W (1972) 'Reprisals Involving Recourse to Armed Force', *American Journal of International Law* 66, 1: 1–36.

Brennan, J. (2011) 'Remarks of John O. Brennan, "Strengthening our Security by Adhering to our Values and Laws"', The White House, Washington, D.C., www.whitehouse.gov/the-press-office/2011/09/16/remarks-john-o-brennan-strengthening-our-security-adhering-our-values-an (accessed 28 August 2013).

Brownlie, I. (1958) 'International Law and the Activities of Armed Bands', *International and Comparative Law Quarterly*, 7, 4: 712–35.

Cassese, A. (1989) 'The International Community's "Legal" Response to Terrorism', *International and Comparative Law Quarterly*, 38, 3: 589–608.

Cassese, A. (2001) 'Terrorism is also Disrupting Some Crucial Legal Categories of International Law', *European Journal of International Law*, 12, 5: 993–1001.

Council of Europe (1950) 'European Convention for the Protection of Human Rights and Fundamental Freedoms', Rome, http://conventions.coe.int/Treaty/en/Treaties/Html/005.htm (accessed 28 August 2013).

Dawn (2011) 'Cameron Munter Summoned to FO over Drone Attack', *Dawn.com* (18 March 2011), Islamabad, http://dawn.com/news/614070/cameron-munter-summoned-to-fo-over-drone-attack (accessed 12 January 2014).

Dinstein, Y. (2005) *War, Aggression and Self-Defence*, Cambridge: Cambridge University Press.

Dorronsoro, G. (2008) 'The Security Council and the Afghan Conflict', in Lowe, V. (ed.) *The United Nations Security Council and War: The Evolution of Thought and Practice since 1945*, Oxford: Oxford University Press, pp. 452–65.

Entous, A., Siobhan Gorman, S., and Perez, E. (2012) 'U.S. Unease over Drone Strikes. Obama Administration Charts Delicate Legal Path Defending Controversial Weapons', *The Wall Street Journal* (26 September 2012), http://online.wsj.com/article/SB10000872396390444100404577641520858011452.html (accessed 28 August 2013).

Finn, J.E. (1991) *Constitutions in Crisis: Political Violence and the Rule of Law*, Oxford: Oxford University Press.

Gazzini, T. (2005) *The Changing Rules on the Use of Force in International Law*, Manchester: Manchester University Press.

Gray, C.S. (2007) *The Implications of Preemptive and Preventive War Doctrines: A Reconsideration.* Strategic Studies Institute. US Army War College: Carlisle, PA, www.strategicstudiesinstitute.army.mil/pdffiles/pub789.pdf (accessed 28 August 2013).

Heymann, P.B. and Kayyem, J. (2004) *Preserving Security and Democratic Freedoms in the War on Terrorism: Final Report of the Long-Term Legal Strategy Project*, Report for Belfer Center for Science and International Affairs. Harvard University: Cambridge, MA. http://belfercenter.hks.harvard.edu/publication/2105/preserving_security_and_democratic_freedoms_in_the_war_on_terrorism.html (accessed 28 August 2013).

Ignatius, D. (2008) 'A Quiet Deal with Pakistan', *Washington Post*, http://articles.washingtonpost.com/2008–11–04/opinions/36914399_1_zardari-government-khalid-habib-pakistani-taliban (accessed 28 August 2013).

International Commission of Jurists (2004) *Declaration on Upholding Human Rights and the Rule of Law in Combating Terrorism*, Berlin: International Commission of Jurists (ICJ).

Jones, S.G. (2008) 'Counterinsurgency in Afghanistan', in *Counterinsurgency Study*,

Santa Monica, CA: RAND Corporation, www.rand.org/content/dam/rand/pubs/monographs/2008/RAND_MG595.pdf (accessed 12 January 2014).

Kaplan, M.A. and Katzenbach, N. (1961) *The Political Foundations of International Law*, New York: Wiley.

Kegley, C.W. and Raymond, G.A. (2003) 'Preventive War and Permissive Normative Order', *International Studies Perspectives*, 4: 385–94.

Koh, H.H. (2010) 'The Obama Administration and International Law', Annual Meeting of the American Society of International Law, Washington, D.C.

Krajewski, M. (2002) 'Selbstverteidigung gegen Bewaffnete Angriffe Nicht-Staatlicher Organisationen – Der 11. September 2001 und seine Folgen', *Archiv des Völkerrechts*, 40, 2: 183–214.

Krajewski, M. (2005) 'Preventive Use of Force and Military Actions against Non-State Actors: Revisiting the Right of Self-Defence', *Baltic Yearbook of International Law*, 5: 1–26.

Kretzmer, D. (2005) 'Targeted Killing of Suspected Terrorists: Extra-Judicial Executions or Legitimate Means of Defence?', *European Journal of International Law*, 16, 2: 171–212.

Levenfeld, B. (1982) 'Israel's Counter-Fedayeen Tactics in Lebanon: Self-Defense and Reprisal under Modern International Law', *Columbia Journal of International Law*, 21: 1–48.

Lobel, J. (1999) 'The Use of Force to Respond to Terrorist Attacks: The Bombing of Sudan and Afghanistan', *Yale Journal of International Law*, 24: 537–58.

Machon, M.J. (2006) *Targeted Killing as an Element of U.S. Foreign Policy in the War on Terror*, School of Advanced Military Studies, Fort Leavenworth, KS: Command and General Staff College.

McCrisken, T. (2013) 'Obama's Drone War', *Survival*, 55, 2: 97–122.

Meisels, T. (2004) 'Targeting Terror', *Social Theory and Practice*, 30, 3: 297–326.

Miller, G., Whitlock C., and Gellman, B. (2013) 'Top-secret U.S. Intelligence Files Show New Levels of Distrust of Pakistan', *Washington Post* (2 September 2013), Washington, D.C., http://articles.washingtonpost.com/2013-09-02/world/41690725_1_intelligence-u-s-spy-agencies-other-adversaries (accessed 20 November 2013).

Murphy, S.D. (2008) *The International Legality of U.S. Military Cross-Border Operations from Afghanistan into Pakistan*, George Washington University Law School Public Law and Legal Theory Working Paper, George Washington University: Washington, D.C., http://papers.ssrn.com/sol3/papers.cfm?abstract_id=1296733 (accessed 28 August 2013).

Myjer, E.P.J. and White, N.D. (2002) 'The Twin Tower Attack: An Unlimited Right to Self-defence?', *Journal of Conflict and Security Law*, 7, 1: 5–17.

New America Foundation (2013) 'The Drone War in Pakistan', in *Analysis*, Washington, D.C., http://natsec.newamerica.net/drones/pakistan/analysis (accessed 12 January 2014).

O'Connell, M.E. (2002) 'The Myth of Preemptive Self-Defense', *ASIL Task Force Papers*, The American Society of International Law: Washington, D.C., www.asil.org/taskforce/oconnell.pdf (accessed 28 August 2013).

OAS (2008) 'Convocation of the Meeting of Consultation of Ministers of Foreign Affairs and Appointment of a Commission', Washington, D.C., www.oas.org/consejo/resolutions/res930.asp (accessed 28 August 2013).

Obama, B.H. (2012) 'Google Hangout Discussion', Youtube, www.youtube.com/watch?v=Nndm5nHzzkg (accessed 12 January 2014).
Pakistan Ministry of Foreign Affairs (2013) 'Pakistan Protests Drone Strike; US CdA summoned', Islamabad, www.mofa.gov.pk/pr-details.php?prID=1264 (accessed 28 August 2013).
Ramsden, M. (2011) 'Targeted Killings and International Human Rights Law: The Case of Anwar Al-Awlaki', *Journal of Conflict and Security Law*, 16, 2: 385–406.
Randelzhofer, A. (2002) 'Commentary to Article 2 (4)', in Simma, B. (ed.) *The Charter of the United Nations: A Commentary*, Oxford: Oxford University Press, pp. 112–35.
Rubin, A.J. and Tavernisse, S. (2008) 'Turkish Planes Strike Iraqi Kurdistan', *New York Times* (5 February 2008), www.nytimes.com/2008/02/05/world/middleeast/05iraq.html?_r=0 (accessed 28 August 2013).
Schachter, O. (1984) 'The Right of States to Use Armed Force', *Michigan Law Review*, 82, 5/6: 1620–46.
Schachter, O. (1989) 'The Lawful Use of Force by a State against Terrorists in Another Country', *Israel Yearbook on Human Rights*, 19: 209–31.
Schmitt, M.N. (2002) 'Counter-Terrorism and the Use of Force in International Law', *Marshall Center Papers* 5, George C. Marshall European Center for Security Studies: Garmisch-Partenkirchen, www.marshallcenter.org/mcpublicweb/MCDocs/files/College/F_Publications/mcPapers/mc-paper_5-en.pdf (accessed 28 August 2013).
Schmitt, M.N. (2012) 'Unmanned Combat Aircraft Systems and International Humanitarian Law: Simplyfying the Oft Benighted Debate', *Boston University International Law Journal*, 30: 595–619.
Schmitz-Elvenich, H.F. (2007) *Targeted Killing. Die Völkerrechtliche Zulässigkeit der Gezielten Tötung von Terroristen im Ausland*, Frankfurt am Main: Peter Lang Verlag.
Shah, S.A. (2010) 'War on Terrorism: Self Defense, Operation Enduring Freedom, and the Legality of U.S. Drone Attacks in Pakistan', *Washington University Global Studies Law Review*, 9: 77–129.
Sofaer, A.D. (1986) 'Terrorism and the Law', *Foreign Affairs*, 64, 5: 901–22.
Stahn, C. (2004) '"Nicaragua is Dead, Long Live Nicaragua" – the Right to Self-defence under Art. 51 UN-Charter and International Terrorism', in Walter, C., Vöneky, S., Röben, V., and Schorkopf, F. (eds) *Terrorism as a Challenge for National and International Law: Security versus Liberty?*, Berlin: Springer, pp. 827–77.
Sterio, M. (2012) 'The United States's Use of Drones in the War on Terror: The (Il)legality of Targeted Killings under International Law', *Case Western Reserve Journal of International Law*, 45, 1/2: 197–214.
Supreme Court of Israel (2006) 'Public Committee against Torture in Israel vs. Government of Israel et al', HCJ 769/02, Jerusalem, http://elyon1.court.gov.il/files_eng/02/690/007/a34/02007690.a34.pdf (accessed 28 August 2013).
U.N. (1999) 'Lettre datée du 12 Juillet 1999, Adressée au Secrétaire Général par le Représentant Permanent de la République Islamique d'Iran auprès de l'Organisation des Nations Unies', *Security Council*, New York, http://daccess-dds-ny.un.org/doc/UNDOC/GEN/N99/205/56/PDF/N9920556.pdf?OpenElement (accessed 28 August 2013).
U.N. (2001) Resolution 1373 (2001), New York, http://daccess-dds-ny.un.org/doc/

UNDOC/GEN/N01/557/43/PDF/N0155743.pdf?OpenElement (accessed 28 August 2013).

U.N. (2002) 'When Force Is Considered, There Is No Substitute for Legitimacy Provided by United Nations, Secretary-General says in General Assembly Address', Department of Public Information, SG/SM/8378 New York, http://domino.un.org/UNISPAL.nsf/eed216406b50bf6485256ce10072f637/4798d1863931fdd285256c36004c3f28?OpenDocument (accessed 28 August 2013).

U.N. (2010) *Report of the Special Rapporteur on Extrajudicial, Summary or Arbitrary Executions, Philip Alston,* Human Rights Council, New York: General Assembly, www2.ohchr.org/english/bodies/hrcouncil/docs/14session/A.HRC.14.24.Add6.pdf (accessed 28 August 2013).

U.N. Economic and Social Council (2004) *Report of the United Nations High Commissioner for Human Rights and Follow-Up to the World Conference on Human Rights. Effective Functioning of Human Rights Mechanisms,* www.unhchr.ch/huridocda/huridoca.nsf/0/fd10a8b6cbcb8e2fc1256db000577bc5?OpenDocument (accessed 28 August 2013).

Walsh, D. (2011) 'Osama bin Laden Killing Prompts US-Pakistan War of Words', *Guardian* (4 May 2011), www.theguardian.com/world/2011/may/04/osama-bin-laden-pakistan-us (accessed 28 August 2013).

Wandscher, C. (2006) *Internationaler Terrorismus und Selbstverteidigungsrecht.* Berlin: Duncker & Humbolt.

Webster, D. (1842) 'Letter to Lord Ashburton', in *The Avalon Project,* New Haven, CT: Yale Law School.

White House (2002) *The National Security Strategy of the United States of America. Strengthen Alliances to Defeat Global Terrorism and Work to Prevent Attacks Against Us and Our Friends,* National Security Council, Washington, D.C., http://georgewbush-whitehouse.archives.gov/nsc/nss/2002/nss3.html (accessed 12 January 2014).

Whitlock, C. (2008) 'In Hunt for Bin Laden, a New Approach', *Washington Post* (10 September 2008), http://articles.washingtonpost.com/2008–09–10/world/36805025_1_al-qaeda-qaeda-senior-pakistani-security/2 (accessed 28 August 2013).

Wikileaks (2011) 'Immunity for Musharraf likely after Zardari's Election as President', http://wikileaks.org/cable/2008/08/08ISLAMABAD2802.html (accessed 17 June 2011).

Wuschka, S. (2011) 'The Use of Combat Drones in Current Conflicts: A Legal Issue or a Political Problem?', *Göttingen Journal of International Law,* 3: 891–905.

Zimmermann, A (2007) 'The Second Lebanon War: Jus ad bellum, jus in bello and the issue of Proportionality', in Bogdandy, A. von and Wolfrum, R. (eds) *Max Planck Yearbook of United Nations Law,* Leiden: Brill Academic, pp. 99–141.

Zimmermann, K. (2013) 'Al Qaeda's African Surge Threatens the U.S.', *The Wall Street Journal* (24 September 2013), http://online.wsj.com/news/articles/SB10001424052702304213904579094961529497766 (accessed 20 November 2013).

8 UAV for R2P?

Exploring the effectiveness and legitimacy of drones

Ina Wiesner

Introduction

The current discourse on unmanned aerial vehicles (UAVs; 'drones') seems paradoxical. Drones are portrayed as the peak of technological advance and at the same time as the harbinger of unregulated war. Borrowing from organisational sociology the antagonistic concepts of effectiveness and institutional legitimacy it is posited that when the effectiveness of drones is discussed the arguments often run in favour of UAVs while the legitimacy of drone operations is framed in a negative context. Drawing on the debate about the Responsibility to Protect, this chapter discusses the potentially positive ramifications of the increasing drone availability for crisis scenarios in which militarily powerful states with casualty-averse societies were reluctant to intervene in the past. It is argued that by dispelling nations' doubts about casualties UAVs could result in a leap of legitimacy by making it easier for states to engage militarily in conflicts where the international community is calling for action.

In recent years a vivid debate has unfolded in public, political, military, and scholarly circles about the use of UAVs in military and covert operations.[1] Contributions to the discussion have covered very diverse aspects – legal, ethical, technological, military-operational, and political (Brunstetter and Braun, 2011; Lange, 2003; Quintana, 2008; Singer, 2009; Sparrow, 2009). As a result, we should rather think of *drone debates* than of *a drone debate.* What is interesting, though, is that within the different debates each discourse is rather homogeneous, whereas between those fields only little exchange of arguments seems to exist. From the perspective of the current debates in the U.S. and in Europe it appears that the (established) political and military actors broadly argue in favour of the purchase and deployment of drones – both for intelligence, surveillance, and reconnaissance (ISR) functions and for offensive combat tasks. Usually, in these circles it is argued that drones are cheaper to purchase than their manned alternatives, require less maintenance work, and are easier to operate, especially in remote locations. In addition, the argument is made that UAVs do not put pilots' lives at risk.

A telling example of such a positive view about UAVs is the series of newspaper articles, parliamentary statements, and conference talks the then German Federal Minister of Defence Thomas de Maizière released between summer 2012 and spring 2013.[2] Against the background of an expiring leasing contract with Rheinmetall for unarmed reconnaissance Heron drones currently deployed to Afghanistan, de Maizière signaled the Ministry's willingness to procure combat drones after 2014, seeking broad public support with statements such as the following:

> UAV offer continuous reconnaissance with a long operating time – much longer than any pilot can stay awake – and surveillance data that are transferred in real time in a way no aircraft could provide. That is why they are technologically useful. They are also not as expensive as aircraft, because fundamental expenses for aircraft go into protecting the pilot; which is redundant for drones. Second, the future of aviation as a whole will be very much shaped by unmanned aerial vehicles of any kind in the next 20, 30, 40 years ... Germany must be part of this future technology. We cannot say we stay with the stagecoach, while everybody else develops the railway.... Third, the introduction of drones also makes sense from a tactical and security policy viewpoint.... Suppose we send a patrol into a dangerous combat situation.... No other tool is as well suited as drones to accompany them, to watch from the air what happens, and then, when our own soldiers are in danger, even to fight without having to call close air support that would only be there after 10, 15 minutes, that does not have a precision capability, and that risks the lives of our soldiers. We do not want that.
>
> (Deutscher Bundestag, 2013)

Prominent in this statement is de Maizière's portrayal of drones as operationally and tactically effective military platforms. Additionally, he stresses the aspect of their efficiency by pointing out that they can perform their tasks in a more cost-efficient way than piloted aircraft. It is interesting to see that he also emphasises quite another aspect of efficiency – the importance of drone technology for a competitive German military research and development base.

Turning to the public and the scholarly debate the focus on the topic as well as the attitude towards drones changes dramatically (Quintana, 2008: 13). The effectiveness or efficiency of drones is hardly ever discussed. Instead a variety of questions focusing on the wider consequences of the technology are raised. Does the deployment of drones violate the fundamental basics of the *jus ad bellum* and the *jus in bello* (Brunstetter and Braun, 2011: 342–55)? Is remote warfare unfair? Does it endanger civilians? Do UAVs deny their human targets the opportunity to surrender? And, more broadly, does the deployment of drones undermine the principle of sovereignty in international relations? Does it lower the thresholds of the military option (Johansson, 2011: 282)? Will it change the legal definition of war (Sharkey, 2014: 183–4)? These and other questions throw a rather critical light on the use of drones. In Germany, for example, the public discussion following the Defence Minister's announcement that he was planning a combat drone purchase was 'oddly removed from the question of actual military requirements and centred on ethical questions, morality, and international law' (International Institute for Strategic Studies, 2013: 100).

Therefore, while it could be said that military and political considerations mostly focus on the positive aspects of the efficiency and effectiveness of drones, the public and scholarly debate is largely concerned with the negative legal and ethical aspects of robotic warfare – that is to say: they focus on the problematic legitimacy of the use of drones.

This chapter explores whether the debate about drones – or, the debates – could benefit from a fresh look at the issue. It will investigate in particular aspects of drone legitimacy that have previously been overlooked. It will examine whether the existence and use of drones could make a difference when it comes to safeguarding populations from crimes against humanity and thus contributing to an increased institutional legitimacy of those states and military organisations that use drones in this particular way. Additionally, the potential benefits of drones for peace-keeping and humanitarian assistance are explored. The chapter thus aims at closing the apparent rift between proponents and opponents of remote warfare.

The chapter starts by briefly introducing the concepts of military effectiveness and institutional legitimacy. It proceeds by exploring how the discourse about drones has been framed so far from the perspective of these concepts. It will be shown that up until the present day some important aspects with regards to the legitimate use of force, in particular with a view towards preventing or stopping crimes against humanity, have not been covered in the discussion. The chapter will close with a discussion of these under-appreciated traits of the UAV topic.

Before proceeding, however, two points regarding the scope of the argument and the fundamental attitude towards technology need clarification. On the first point, this chapter draws partly on the current debate about the responsibility of states and the international community to help protect populations from crimes against humanity, a debate that is labelled as the Responsibility to Protect (R2P). The emergence of the R2P concept was a direct result of the genocide in Rwanda (1994) and the massacre at Srebrenica (1995). Sadly, the debate about R2P is still active due to incidents of systematic and large-scale violence against populations as most recently in Libya (2011) and Syria (2011 to the present). Current situations undeniably require the attention of politicians and scholars alike, especially with a view to how to close the apparent gap between the intention of the international community to protect the lives of individuals and case-specific strategic considerations of the powerful states in the UN Security Council. This chapter, however, does not engage in a discussion about the state of the R2P concept after its somewhat ambiguous appearance in UN Security Council resolutions 1970 and 1973 on Libya (2011) – in which the Libyan authorities' but not the international community's responsibility to protect the Libyan population was mentioned – and its subsequent absence from any resolution on the Syrian crisis. Instead, this chapter discusses the relationship between military technology and military engagement on a more theoretical level.

Second, any attempt to address the influence of technology on politics runs the risk of arguing in a deterministic way. Technological determinism posits that it is the technological circumstances that shape social outcomes and not human will (Bimber, 1994: 82–4). This chapter explores how the drone capability at the disposal of military organisations might affect states' choices in becoming militarily engaged in conflicts and crises elsewhere. It is not, however, argued here that drone technology can resolve international conflicts or that drones are an ultimately *good* technology – just as they are not *evil* by definition. They are purchased and deployed through purposeful acts of political and military actors. The idea elaborated in the following sections – that the possession of UAVs might ease decision making about military engagement in conflicts in which populations are targeted and thus lead to an increased international legitimacy for those states using them – should not imply an automatism in the national and international decision making on military engagements. The introduction of drones into the military portfolio provides just one factor among many for politicians to take into account when deciding about the set-up of and participation in peace-enforcement or peace-keeping missions. Yet, even though the existence of UAVs might not determine national decision making on these matters, it is nevertheless likely to have an impact in the military pre-planning and in political negotiations. To sum up this final word of caution, drones undoubtedly *have politics*,[3] but they do not *make* them.

Effectiveness and legitimacy of military organisations

In organisational sociology the concepts of effectiveness[4] and legitimacy (sometimes also referred to as the logic of consequence and the logic of appropriateness (March and Olsen, 2006)) represent two sides of one coin when discussing the structuring and functioning of organisations (Friedberg, 2001; Meyer and Rowan, 1991). The fundamental question is what determines the outcome of organisational action. Organisations could be viewed as rational maximisers of effectiveness that do what they do to achieve their aims preferably in a cost-effective way. At the same time organisations could be understood as being subject to external and internal constraints on their action – be that laws, norms, expectations or even (undisputed) routines by their staff. Ideally those constraints first do not contradict each other and second do not jeopardise the effective functioning of organisations. However, organisations are often in a constant struggle to meet – sometimes conflicting – demands from their social environments and to achieve their own aims at the same time (Boxenbaum and Jonsson, 2008; Pfeffer and Salancik, 1978).

The concepts of effectiveness and legitimacy also apply to military organisations; both logics are important drivers of military action. Military effectiveness, to start with, can be understood in four dimensions (Millett *et al.*, 1986): (i) politically, the military competes with other public institutions for resources: (ii) strategically, the military needs to translate political goals into a military plan; (iii) operational effectiveness evolves from meaningful analysis, planning, preparation, and conduct of military campaigns, whereas (iv) tactical effectiveness refers to military techniques used in combat situations.

Like any other organisation, however, military organisations are not mere rational optimisers of efficiency and effectiveness. They also need to maintain and gain legitimacy. To date there is not an agreed upon meaning of the word legitimacy. The understanding of the word depends on whether one is interested in questions of, for example, political authority, the morality and legality of action – or, as discussed in this chapter – organisational behaviour. For the latter, three broad perspectives have been identified (Stryber, 2001: 8701): legitimacy, first, as a normative mechanism (adherence to rules as a reflection of the internalisation of beliefs), second, as an instrumental mechanism (adherence as a strategy to maintain external support), and, third, as a constitutive mechanism (adherence as the re-institutionalisation of the system). For the purpose of analysing the distribution of arguments in drone discourse the second perspective – the instrumental – applies. In this view, legitimacy can be understood as the social acceptability and credibility that organisations require in order to "survive and thrive in their social environment" (Scott, 2001: 237). In this instrumental understanding of the concept legitimacy is comparable with a currency that is paid when the organisation acts in accordance with what is expected from its institutional environment. These expectations are in themselves value-neutral. This is an important difference from other understandings of legitimacy that carry ethical or legal notions and that are often employed in the current drone debate (Kennedy, 2013).

For military organisations institutional incentives and constraints to action stem from the three levels they depend upon: the internal workforce, the national level (e.g. the political system, society, industry) and the international level (e.g. the international legal system, international organisations, alliances). These levels constitute the institutional environment for military organisations and confront or provide the military organisation with demands, incentives, support, resources and, ultimately, (institutional) legitimacy.

Often military action is less effective than it could be due to existing constraints from the environment. Most profoundly, the majority of states have agreed to international norms and laws, such as the general principle of proportionality or more specific norms like the nuclear, chemical, and biological weapons taboos that govern the use of force. Clearly, these laws and norms limit states in their purchase and deployment of otherwise effective weapons. In terms of institutional legitimacy the laws of war are the strongest social constraint to the maximisation of the utility of military means by states.

But norms also influence the specific choice of assets even within the range of legally or morally accepted military capabilities. The military operations in Kosovo (1999) and Iraq (2003–09), for example, have spurred the debate on whether in some situations the deployment of ground troops (or, in the case of Kosovo, the threat thereof) was not more effective than the deployment of airpower. Yet governments in western societies prefer to "go light" by reducing the number of troops deployed to foreign theatres in order to avoid friendly casualties, as casualty aversion has a growing impact on societies' acceptance of the use of (military) force. From the military organisation's perspective a society's casualty aversion presents a constraint to its actions as much as the laws of war, albeit the latter stems from a different level of the organisation's institutional environment (the international level) and shows a much higher degree of institutionalisation than the former.

Constraints on organisational action might also stem from within the organisation, from the workforce. In many European states the average standing time for soldiers in missions abroad varies between four to six months whereas in the U.S. military troops stay in mission up until 12 months. In terms of continuity of work the latter option is probably more effective yet in some societies the absence of fathers and mothers, wives and husbands for more than six months simply does not seem appropriate.

In sum, military organisations, on the one hand, operate in highly technical environments and are subject to considerations of effectiveness in their daily functioning and especially while on mission. Yet, they are also dependent on internal, societal, political, and international support in order to legitimise their purpose, their missions, and rules of engagement (Wiesner, 2013: 65).

The current debate about drones has covered aspects of effectiveness and legitimacy alike. But it seems that supporters of UAVs mostly stress the technical benefits of the systems whereas opponents claim that drones violate fundamental principles of the laws of war. Table 8.1 summarises the foci in the debate about UAVs (Brunstetter and Braun, 2011; Byman, 2013; Cronin, 2013). The arguments are presented along the effectiveness/legitimacy divide.

Table 8.1 Arguments in the current drone debate[5]

	Proponents of UAV	*Opponents of UAV*
Effectiveness		
Political	UAVs more economical than aircraft	Benefits outweighed by the wider foreign policy costs
Strategic	Drones offering technological advancements against conventional and unconventional enemies	High-tech weapons less useful in asymmetric scenarios
Operational/ Tactical	Depending on the particular UAV system: long duration, all weather capability, loitering, ISR gathering, combat tasks, easily deployable, low maintenance requirements, part of a network-centric military	Command of the air necessary to operate UAV; Problems: network resilience, malfunctioning of UAV, information overload of the operators
Legitimacy		
Organisational	Remote operations (protecting soldiers)	Disconnecting soldiers from their tasks
Societal/Political (national)	Remote operations (reducing own casualties)	
International		Violation of *jus ad bellum* Violation of *jus in bello*

The distribution of arguments suggests that issues of legitimacy are mostly limited to discussing the positive effects of remotely operated systems on the physical integrity of pilots. This argument touches issues of legitimacy on two levels, the organisational and the societal/political. Military organisations can increase their internal legitimacy vis-à-vis their own or coalition troops by putting pilots out of harm's way using drone technology (employer's duty of care argument). Likewise the remoteness of military operations might become an important pro-argument (casualty-avoidance argument) when foreign military engagement is debated in casualty-averse western societies (Dowd, 2013: 8; Vennesson, 2011) thus leading to the maintenance or increase of institutional legitimacy of the military organisation.

Opponents of UAVs have little or nothing to reply to the claims of effectiveness and efficiency made by the proponents.[6] On a very general level military experts with a critical view of the current wave of information-based military revolution have repeatedly raised concerns regarding the technocentrism (Kaufman, 2002) and the overall usefulness of networked militaries in conventional battles (Biddle, 2003), in asymmetric conflicts (Hammes, 2004: 205), as well as in the fight against international terrorist groups (Cronin, 2013). Furthermore, critics question the resilience of communication networks (Kaufman, 2002: 22), the substitution of human intelligence by technology (Murray and Scales, 2003: 182), the physical stress (Sparrow, 2009) and information overload (Brown, 2001) of military personnel, the actual cost efficiency of the new platforms (Joshi and Stein, 2013: 55–7), their operational reliability (Dowd, 2013: 15) and, finally, the requirement for the command of the air for today's high-tech operations (Cronin, 2013: 53).

Compared to the plethora of praises for the benefits of drones these counterarguments, however, while far from being unjustified, appear to be too few to have an impact on the debate. Arguments against UAVs are only rarely made on the grounds of the technology's effectiveness or leave the proponents unimpressed. Instead, drones have met their strongest criticism on the level of legitimacy. But opponents tend to emphasise an aspect of legitimacy different from the proponents. It is not so much the impact of military drones on society or on the military organisation – although some concerns have been raised about the potential alienation of soldiers from their tasks – that seems to be worrisome in the eyes of objectors, but the consequences of remote weapons for the way war has been contained in the past. Scholars of international law and international relations discuss the consequences of the deployment of drones, especially with a view to extraterritorial killings – the harm done to non-combatants, the effects on public opinion in foreign states, the opening of a Pandora's box of systematically violating sovereign borders and drones in the light of the ethics of war (Amstutz, 2013; Brunstetter and Braun, 2011; Etzioni, 2013).

Supporters of the purchase and use of drones hardly engage in this discussion and if they do it is as a reaction to claims made by opponents. Concerning the issue of lacking discrimination, it has been argued recently that UAVs cause fewer – not more – civilian casualties than alternative means of war (Byman, 2013: 32). When faced with the strong opposition to his plan of procuring combat drones for Germany, German Minister of Defence de Maizière argued in a nation-wide newspaper in August 2013 that drones were not different from any other type of weapon. They were, in fact, an 'ethically neutral' weapon.[7] Earlier that year de Maizière had elaborated on this view:[8]

> Whether you shoot a torpedo from a submarine, fire a guided missile from the ground, fire a missile from an aircraft to the ground or deploy and fire an armed drone, it is always the same rules, and therefore also the same legal rules.

De Maizière's statements about the UAV illustrate the peculiar division of arguments in the current drone debate. While there is much talk about the technological advantages and some arguments against claims that drones violate principles of the laws of war, only few proponents elaborate on the potential benefits of drones in terms of maintaining or increasing institutional legitimacy from the international level. So far, arguments that would fall in the lower left cell in Table 8.1 are only rarely made (Beauchamp and Savulescu, 2013). However, both the scholarly and the public drone debate might profit from a systematic discussion of the impact of UAVs on international security.

A key to understanding a more comprehensive account of this impact is to lose the narrow focus on extrajudicial killings and offensive military operations that understandably dominates today's view on the subject and to include in the discussion the use of drones in other areas of conflict and conflict resolution, such as in humanitarian assistance operations, stabilisation missions, and military interventions (Johnson, 2013; Whetham and Strawser, 2013). Additionally, when debating the effectiveness of UAVs the view should not be limited to offensive combat operations but also explore the support functions that drones can offer to military organisations in crisis management operations.

The following sections explore how the existence of UAVs in the militaries' toolkits could affect the security of populations in war-torn countries. It will be argued that drones can provide better monitoring in peacekeeping operations and in humanitarian assistance missions. Moreover, drones might positively impact the world's pledge to come to the rescue of those suffering from war crimes, genocide, crimes against humanity, and ethnic cleansing by encouraging militarily powerful states to take more seriously the R2P concept.

Drones for good?

Effectiveness of drones in peace-keeping operations

In terms of effectiveness and efficiency the deployment of drones offers a number of functions: most importantly, drones perform a variety of ISR tasks unmatched by any other available manned system. Surely, ISR is important in battle but also in other scenarios short of war. Reliable intelligence and steady surveillance could make a difference in peace-keeping and peace enforcement as well as in non-military routines such as conflict prevention, in post-conflict confidence building, and even in humanitarian assistance missions. Currently, the UN assesses the use of drones to enhance its situational awareness.[9] In its resolution 2098 (2013) the UN Security Council authorised the United Nations Organization Stabilization Mission in the Democratic Republic of the Congo (MONUSCO) to use UAVs for monitoring the implementation of the arms embargo. If UAVs successfully pass the trial use in MONUSCO the technology could become a contribution to both UN peace-keeping and the UN's early warning capacity. Certainly, drones will not replace human intelligence gathering but they might make a difference by complementing pieces of information acquired. In that way UAVs might help systems of collective security such as the UN to get a better picture of conflict situations.

A special case could be the engagement of drones in conflict zones in which violence has already broken out but where the UN Security Council is unable to reach an agreement to use 'all necessary means'. In August 2013, David Whetham and Bradley Strawser, for example, suggested deploying unarmed UAVs to Syria to monitor the situation on the ground, especially with a view to the use of chemical weapons (Whetham and Strawser, 2013). Drones could offer a way at least to bear witness to atrocities and thus provide the Security Council, the UN, and the world with information about what is happening on the ground. Not only would the UN be able to obtain 'neutral' information about the situation, relying neither on information from government forces nor video footage of the Free Syrian Army, and without worrying about the safety of observation teams (Whetham and Strawser, 2013); the systematic deployment of surveillance drones could deter further war crimes if backed by credible threats of punishment (for example, the passing of evidence to the International Criminal Court). Likewise, formerly warring parties might agree to let their actions be monitored by drones as a way to re-build confidence after cease-fire agreements. In a similar way, drones could routinely carry out military arms control and verification tasks (Richter, 2011: 106).

Finally, the deployment of drones could matter in humanitarian assistance situations, especially for the protection of refugee camps. Considering the reasons for the phenomenon of child soldiers Vera Achvarina and Simon Reich made the case that the typical arguments for the high number of child soldiers found in the literature – poverty levels, orphan rates, the availability of small arms – were insufficient to account for variations in the ratio of child soldiers in different conflicts (Achvarina and Reich, 2006). The authors offered an alternative explanation – 'the degree to which children are protected in refugee camps is the primary determinant of child soldier recruitment rates' (ibid.: 132). Backed with a credible threat of a quick reaction team on the ground drones could monitor refugee camps or other civilian spaces in war-torn regions. By that they would not only add some ISR capability to protection forces but also a show-of-force element, which potentially deters perpetrators of violence. In the current debate the negative psychological effects of the presence of armed drones in rural areas of Pakistan are widely discussed.[10] But could the sound of a UAV routinely patrolling UN-protected sites not lead to a feeling of security for those who have sought protection from UN agencies?

In sum, an increased availability of drones for UN peace-keeping missions is likely to increase the organisation's military effectiveness. But when debating the advantages of UAVs it is mostly the platform's utility for robust military missions that is discussed – as the parliamentary statement of then German Minister of Defence exemplifies. Yet, this unnecessary reduction of the platform's actual value for the whole mission spectrum hinders a balanced appreciation of its merits. To be sure, the deployment of UAVs for peace-keeping will not alter the logic of these operations. Peace-keeping will continue to rely on the assignment of high numbers of soldiers, on human intelligence, and on interpersonal contacts between peace-keepers and conflicting parties and civilians. Like the employment of any other technology the use of drones might additionally lead to unintended and/or undesired consequences and might also be subject to human error (for example, data misinterpretation) or even wrongful application. Yet the leap in surveillance capability which is offered by drones could increase the effectiveness of peace-keeping operations and thus should be taken into account when debating drone capability.

Drones and aspects of legitimacy

Apart from increasing the effectiveness of monitoring, peace-keeping, and humanitarian assistance, the deployment of drones could also enhance the legitimacy *of* UN missions and other coalitions to stop violence against civilian populations as well as the legitimacy *in* such missions. The deployment of ISR-UAVs could make decision making within UN missions more credible by offering additional channels for information gathering. 'UN missions must be impartial', argues Adrian Johnson, 'bolstering their organic surveillance capability can reduce their reliance on potentially unreliable eye-witness reports and thinly stretched patrols, providing a more-rounded basket of intelligence' (Johnson, 2013). Furthermore, when the security situation calls for international action the availability of UAVs – both ISR and combat drones – could ease the reluctance of western political leaders to engage in peace-keeping or even in peace-enforcement missions. If drones can perform some military functions that have previously been executed by soldiers then war-weary western states might be more willing to allocate troops when those gruesome situations occur that make the use of military means seem unavoidable. This argument is at the heart of the *jus ad bellum* debate and so far has been used by the opponents of UAVs rather than by their proponents.

The common argument against (combat) drones is that by their reduced risk to soldiers' lives, their reduced procurement and maintenance costs, and a supposedly lower barrier for gathering public and political support for deploying them, powerful states would more easily resort to the military option, thereby neglecting non-military means of conflict-resolution. But again, could a shift of the argument not bring about some interesting contributions to the debate? From the perspective of victims and potential victims of crimes against humanity the increasing availability of drones could to some extent ease the reluctance of the states to come to their help. In particular, it might be interesting to connect the current discourse about drones with the on-going debate about the commitment of the international community to intervene in cases of systematic human rights violations, the debate about the responsibility to protect (R2P).

In brief, R2P describes the duty of the state to protect its citizens from genocide, war crimes, crimes against humanity, and ethnic cleansing.[11] If the state is unable or unwilling to do so, the international community has in theory committed itself to act in order to protect the population from these crimes – and this might even mean resorting to military means. By that, the concept challenges the principles of state sovereignty and non-interference (Evans, 2008). In addition, R2P calls into question the previous understanding of humanitarian intervention as a "right" of intervention and posits some kind of obligation to react in the face of systematic violations of basic human rights.

However this is not straightforward. Current academic and political discourse is mainly limited to discussing the repercussions of the military option for the non-intervention norm in international law. It is contested how far the concept of responsibility could be stretched and if there is such a thing as a duty or obligation to intervene or not (Arbour, 2008; Bain, 2010). Likewise, the concept's viability as a norm, or even as an 'emerging norm' (UN, 2004: 66), is under scrutiny (Hehir, 2013: 37), because, as some argue, 'the vast majority of states simply does not want to be legally bound to save strangers in remote regions of the world' (Reinold, 2010: 57). Rather than exploring the circumstances of consistent military international responses to mass atrocities, the UN tried to foster the aspect of *preventing* the above mentioned crimes in the first place by appointing Special Advisers on the Prevention of Genocide (2004)[12] and on the Responsibility to Protect (2007),[13] by establishing a joint early warning and risk assessment capacity[14] and by collaborating with regional and sub-regional organizations such as the European Union (EU), the African Union (AU), and the Economic Community of West African States (ECOWAS) on the matter of R2P.

Applying the concept of institutional legitimacy it could be argued that even though R2P is at the most only weakly institutionalised, the emergence of the concept at the international level already presents states with an additional set of demands – to help stop crimes against humanity regardless of their strategic interests being touched – that they have to take into consideration when deciding about military engagement, or non-engagement. This additional set of demands, however, will not determine state action in the future.

It has been pointed out above that, for social organisations, demands from the environment are often conflicting and clearly the still nascent concept of R2P seems largely inconsistent with strong and established norms such as non-interference. Adding – at the national level – the general war-weariness of the U.S. after the Iraq War and NATO at the end of a decade-long engagement in Afghanistan it could be assumed that in the future western states will be less willing to devote military resources and soldiers to stop crimes against humanity should they occur (Abramowitz, 2013) or that they will try to minimise the costs of such interventions (Murray, 2013: 31).

How, then, could UAVs resonate with R2P and how could they contribute to increasing the institutional legitimacy of military organisations? The answer is twofold. First, ISR-UAVs at the disposal of the UN could significantly increase the organisation's ability to enhance its early warning capacities, thereby contributing to the prevention aspect of R2P. Second, the increased availability of unmanned systems might in turn lessen the potentially growing future reluctance of casualty-averse western governments to deploy forces in conflict regions. By dispelling national fears of own casualties in peace-enforcement operations the possibility of UAV deployment might contribute to a shift in national political decision making on the participation in such operations. For the respective state and military organisation this could lead to an increase in institutional legitimacy stemming from the international level.

Discussing the role of UAVs in the Libyan intervention Whetham (2013: 28) argued that

> [t]he fact that no Western personnel lost their lives during Operation Unified Protector was an added bonus. However, the next time that a pilot participating in a just war is lost in action while carrying out a task that could have been undertaken just as effectively by a remotely piloted drone may well prompt a more fundamental shift in public perceptions of what is acceptable or legitimate.

Still, uneasiness exists about the impact of drone availability on national decision making about foreign policy choices. Critics fear that drones might 'encourage countries to act on just cause with an ease that is potentially worrisome' (Brunstetter and Braun, 2011: 339). Therefore it is important to point out that the argument made in this chapter – that drones might matter in stopping violence against populations – can only hold when states adhere to international law when engaging in areas of conflict. Given the presence of a just and worthy cause, and given that circumstances emerge that call for an international military response, the availability of unmanned systems might be one relevant factor for states to overcome the habitual reluctance to become engaged. Applying the language of (instrumental) organisational legitimacy the availability of drones might make it easier for the politico-military organisation to meet demands from its environment for the protection of the innocent.

Does the availability of drones, then, *determine* a state's course of action when faced with a situation of systematic crimes against humanity elsewhere? Of course, it does not. Demands from the institutional environment are manifold and often conflicting. Moreover, against the background of history and current political circumstances military organisations and their political leaderships vary in their dependence on different parts of the social environment. They have to weigh environmental demands and incentives against each other, extrapolate potential political consequences, and ultimately they have to make a choice for or against military commitment. But by lowering the risk to military personnel drones might alter some parameters in the politico-military organisations' deliberation processes about military engagement.

Conclusions

The rise of unmanned aerial systems undoubtedly affects the way militaries operate. According to UAV proponents militaries can do more with the new technology, they can do it better, and supposedly they can do it in a more cost-efficient way. Yet the use of drones by military organisations also affects the way the military interacts with its institutional environment, notably its internal workforce, the societal and political spheres at the national level as well as the international level. From these levels military organisations receive institutional legitimacy, which they require to justify their existence and actions. While proponents of UAVs stress the fact that putting the pilots out of harm's way resonates with demands from both the organisational as well as the national level in casualty-averse societies – and as such presents an increase in institutional legitimacy – critics of the new technology perceive the consequences of UAV deployment in a different light.

By arguing that the use of UAVs violates the laws governing the use of force both before and during war they implicitly claim that the institutional legitimacy of those military organisations that use drones does in fact decrease. In the current debate demands from the international environment are mostly depicted as being constraining in nature. Most fundamentally, questions are raised concerning the compatibility of UAV uses with principles of the *jus ad bellum* and the *jus in bello*. As justified as many of those concerns about the legal implications of UAVs uses, for example in cases of targeted killings, might be, it should also be noted that demands from the international environment are not always constraining in nature but could as well be encouraging and instigating. This is, for example, the case when the UN Security Council calls for military action under Chapter VII of the UN Charter, especially in cases of crimes against humanity when the strategic interests of the world powers are not at stake.

In this chapter it has been argued that – because of a significantly reduced risk to their own military personnel – states might be more willing to contribute to such missions if their military organisations have been equipped with a drone capability. The UN-mandated prevention or stopping of genocide and other crimes against humanity might, in turn, lead to an increase of international legitimacy for those states that engage in such operations.

The aim of this chapter was to feed into the current scholarly UAV debate a more balanced perspective on the legitimacy of drones, first, by offering a less value-laden but instrumental concept of legitimacy and, second, by exploring the potentially positive ramifications of UAV deployment for international security.

Moreover, the argument made in this chapter might also be important for public discourse on the subject. Like other social organisations, militaries (or, to be more specific, politico-military complexes) are constantly weighing alternative options when deciding about courses of action. When a new technology or operational concept is implemented it is often adjusted so as to fit the existing constraints stemming from the militaries' institutional environment (Wiesner, 2013: 58–65, 133–5). Differences in concept adoption between military organisations can also be explained by differences in environmental demands as well as by differences in the dependence on, and thus their response to, the institutional environment. Even without a systematic assessment of the particular circumstances it can be noted that, for example, the U.S. military currently plays UAVs to their full strength whereas in Germany the public debate about drones – let alone their actual purchase for combat tasks – is highly contested. But it would be erroneous to assume that the relationship between social organisations such as militaries and their institutional environment is unidirectional. The armed forces are embedded in a thick net of mutually influencing relationships with their environment. To some extent social organisations impact on their environment (Pfeffer and Salancik, 2003: ch. 4) and also have an influence on discourses as well as norm setting.

It might thus be important also for the politico-military level to explore the uses of unmanned aerial vehicles for peace-keeping operations better and also to engage in an honest and transparent discussion about the positive and negative effects of remote ISR and combat capabilities, for example, for the containment of crimes against humanity. Such a broadening of the drone issue might revive and refocus the necessary public debate, especially in countries such as Germany in which the societal level often presents a constraining factor in the German armed forces' institutional environment and in which the public is averse to foreign military deployments and the purchase of 'killer-drones'.[15] Instead of adopting a merely defensive posture in the drone discourse by refuting the concerns of drone critics by arguing that nothing actually changes with the purchase of drones, politicians and scholars alike should also proactively explore the potential benefits and limitations of unmanned systems for international security, especially in those cases when non-military means are exhausted and the military option seems the only way to stop violence.

Notes

1 In preparing this chapter, the author has benefitted from comments provided by Mike Aaronson, Tom Dyson, Anthony Seaboyer, Dirk Schuchardt, Steffen Kraft, and Hendrik Jaeschke. This chapter reflects the personal opinion of the author.

2 www.dw.de/germany-plans-to-buy-armed-combat-drones/a-16272847, www.welt.de/politik/deutschland/article108473948/De-Maiziere-wirbt-fuer-Einsatz-bewaffneter-Drohnen.html, http://augengeradeaus.net/tag/militarbischof (accessed 10 December 2013).

3 On the politics of technological artefacts, see Langdon Winner's essay on the subject (1986: 19–39).

4 For the sake of a parsimonious argument, the notion of efficiency will be subsumed under the term effectiveness throughout the chapter. Millett *et al.* (1986: 37) argued in a similar way: 'Military effectiveness is the process by which armed forces convert resources into fighting power. A fully effective military is one that derives maximum combat power from the resources physically and politically available. Effectiveness thus incorporates some notion of efficiency.'

5 I am grateful to Mike Aaronson and Dirk Schuchardt for critically discussing and amending this compilation of arguments.

6 Two notable exceptions are Dowd (2013), who discusses the risks of drone wars for the US, and Joshi and Stein (2013), who discuss effectiveness and efficiency of drone purchases for non-US countries.

7 www.welt.de/politik/deutschland/article108473948/De-Maiziere-wirbt-fuer-Einsatz-bewaffneter-Drohnen.html (accessed 10 December 2013).

8 Deutscher Bundestag, plenary protocol 17/219 page 27111, own translation.

9 See Letter dated 27 December 2012 from the Secretary-General addressed to the President of the Security Council (S/2013/43) and Letter dated 22 January 2013 from the President of the Security Council addressed to the Secretary-General (S/2013/44). Also see Johnson (2013).

10 The Stanford International Human Rights and Conflict Resolution Clinic and the Global Justice Clinic at NYU School of Law have launched a website looking at this issue: www.livingunderdrones.org (accessed 10 December 2013).

11 Resolution adopted by the General Assembly (2005 World Summit Outcome), A/RES/60/1, www.un.org/en/preventgenocide/adviser/pdf/World%20Summit%20Outcome%20Document.pdf#page=30 (accessed 10 December 2013).

12 Exchange of letters between the Secretary-General and the President of the Security Council, S/2004/567 and S/2004/568.

13 Exchange of letters between the Secretary-General and the President of the Security Council, S/2007/721 and S/2007/722.

14 www.un.org/en/preventgenocide/adviser/country_situations.shtml (accessed 10 December 2013).

15 www.taz.de/!110843/ (accessed 10 December 2013).

References

Abramowitz, M. (2013) 'Does the United States have a "responsibility to protect" the Syrian people?', *Washington Post* (Online Edition), http://articles.washingtonpost.com/2013-09-06/opinions/41827796_1_chemical-weapons-united-nations-crimes (accessed 6 September 2013).

Achvarina, V. and Reich, S.F. (2006) 'No Place to Hide: Refugees, Displaced Persons, and the Recruitment of Child Soldiers', *International Security*, 31, 1: 127–64.

Amstutz, M.R. (2013) *International Ethics: Concepts, Theories, and Cases in Global Politics*, Lanham: Rowman & Littlefield.

Arbour, L. (2008) 'The Responsibility to Protect as a Duty of Care in International Law and Practice', *Review of International Studies* 34, 03: 445–58, http://dx.doi.org/10.1017/S0260210508008115 (accessed 12 January 2014).

Bain, W. (2010) 'Responsibility and Obligation in the "Responsibility to Protect"', *Review of International Studies,* Supplement S1, 36: 25–46, http://dx.doi.org/10.1017/S0260210511000106 (accessed 12 January 2014).

Beauchamp, Z. and Savulescu, J. (2013) 'Robot Guardians: Teleoperated Combat Vehicles in Humanitarian Military Intervention', in Strawser, B.J. (ed.), *Killing by Remote Control: The Ethics of an Unmanned Military*, pp. 106–25.

Bellamy, A.J. (2005) 'Is the War on Terror Just?', *International Relations* 19, 3: 275–96, http://ire.sagepub.com/content/19/3/275.abstract (accessed 12 January 2014).

Biddle, S.D. (2003) 'Afghanistan and the Future of Warfare', *Foreign Affairs,* 82, 2: 31–46.

Bimber, B. (1994) 'Three Faces of Technological Determinism', in Smith, M.R. and Marx, L., *Does Technology Drive History? The Dilemma of Technological Determinism,* Cambridge, MA: MIT Press, pp. 79–100.

Boxenbaum, E. and Jonsson, S. (2008) 'Isomorphism, Diffusion and Decoupling', in Greenwood, R., Oliver, C., Sahlin, K., and Suddaby, R. (eds), *The SAGE Handbook of Organizational Institutionalism,* Los Angeles/London, SAGE, pp. 78–98.

Brown, F.J. (2001) 'Tactical Situational Awareness: The Human Challenge', in Pfaltzgraff, R.L.J. and Shultz, R.H. Jr. (eds), *War in the Information Age,* Washington/London: Brassey's, pp. 99–174.

Brunstetter, D. and Braun, M. (2011) 'The Implications of Drones on the Just War Tradition', *Ethics and International Affairs,* 25, 3: 337–58.

Byman, D. (2013) 'Why Drones Work: The Case for Washington's Weapon of Choice', *Foreign Affairs* (July/August 2013): 32–43.

Cronin, A.K. (2013) 'Why Drones Fail: When Tactics Drive Strategy', *Foreign Affairs* (July/August 2013): 44–54.

Deutscher Bundestag (2013) 31 January, Parliamentary Debate, Plenary Protocol 17/219 pp. 27110–27111, http://dipbt.bundestag.de/extrakt/ba/WP17/506/50645.html, authors translation.

Dowd, A.W. (2013) 'Drone Wars: Risks and Warnings', *Parameters* (US Army War College Quarterly) 42, 4/43, 1: 7–23.

Etzioni, A. (2013) 'The Great Drone Debate', *Military Review,* 93, 2: 2–13.

Evans, G.J. (2008) *The Responsibility to Protect: Ending Mass Atrocity Crimes Once and for All,* Washington, D.C.: Brookings Institution Press.

Friedberg, E. (2001) 'The Sociology of Organizations', in Smelser, N.J. and Baltes, P.B. (eds), *International Encyclopedia of the Social and Behavioural Sciences,* Oxford/New York: Elsevier.

Hammes, T.X. (2004) *The Sling and the Stone,* St. Paul, US: Zenith Press.

Hehir, A. (2013) 'The Responsibility to Protect as the Apotheosis of Liberal Theory', in Hehir, A. and Murray, R. (eds), *Libya: The Responsibility to Protect and the Future of Humanitarian Intervention,* Basingstoke/New York: Palgrave, pp. 34–57.

International Institute for Strategic Studies (2013) Chapter Four: 'Europe'. *The Military Balance,* 113, 1: 89–198.

Johansson, L. (2011) 'Is It Morally Right to Use Unmanned Aerial Vehicles (UAVs) in War?', *Philosophy and Technology*, 24, 3: 279–91.

Johnson, A. (2013) 'Drones in the Service of Peace: Unmanned Aircraft in UN Deployments', *RUSI Analysis* (18 February 2013), www.rusi.org/analysis/commentary/ref:C51223460690F6 (accessed 12 January 2014).

Joshi, S. and Stein, A. (2013) 'Emerging Drone Nations', *Survival*, 55: 5.

Kaufman, A. (2002) 'Be Careful What You Wish For: The Dangers of Fighting With a Network-Centric Military', *Journal of Battlefield Technology*, 5, 2: 20–3.

Kennedy, G. (2013) 'Drones: Legitimacy and Anti-Americanism', *Parameters* (US Army War College Quarterly) 42, 4/43, 1: 25–8.

Lange, S. (2003) 'Flugroboter statt Bemannter Militärflugzeuge?', *SWP Studie* (SWP Berlin), www.swp-berlin.org/common/get_document.php?asset_id=187 (accessed 12 January 2014).

Mahnken, T.G. (2008) *Technology and the American Way of War*, New York: Columbia University Press.

March, J.G. and Olsen, J.P. (2006) 'The Logic of Appropriateness', in Moran, M., Rein, M. and Goodin, R.E. (eds), *The Oxford Handbook of Public Policy*, Oxford: Oxford University Press, pp. 689–708.

Meyer, J.W. and Rowan, B. (1991) 'Institutionalized Organizations: Formal Structure as Myth and Ceremony', in Powell, W.W. and DiMaggio, P.J. (eds) *The New Institutionalism in Organizational Analysis*, Chicago: Chicago University Press, pp. 41–62.

Millett, A.R., Murray, W., and Watman, K.H. (1986) 'The Effectiveness of Military Organizations', *International Security*, 11, 1, 37–71.

Murray, R.W. (2013) 'Humanitarianism, Responsibility or Rationality? Evaluating Intervention as a Strategy', in Hehir, A. and Murray, R. (eds), *Libya: The Responsibility to Protect and the Future of Humanitarian Intervention*, Basingstoke/New York: Palgrave, pp. 15–33.

Murray, W. and Scales, R.H. (2003) *The Iraq War: A Military History*, Cambridge, MA: Belknap Press of Harvard University Press.

Pfeffer, J. and Salancik, G.R. (1978) *The External Control of Organizations: A Resource Dependence Perspective*, rev. edn 2003, Stanford, CA: Stanford Business Books.

Quintana, E. (2008) 'The Ethics and Legal Implications of Military Unmanned Vehicles', RUSI Occasional Paper, www.uvs-info.com/phocadownload/03_5dc_Relevant_Information/Block-2_Ethics_RUSI_Ethics%20and%20Legal%20Implications%20of%20Military%20Unmanned%20Vehicles%20(RUSI).pdf (accessed 16 January 2014).

Reinold, T. (2010) 'The Responsibility to Protect: Much Ado About Nothing?', *Review of International Studies*, 36, Supplement S1: 55–78, http://dx.doi.org/10.1017/S0260210510000446 (accessed 12 January 2014).

Richter, W. (2011) 'Kampfdrohnen versus Völkerrecht? Zum "Drohnenkrieg" in Afghanistan und Pakistan', *Humanitäres Völkerrecht*, 24, 2: 105–12.

Scott, W.R. (2001) *Institutions and Organizations: Foundations for Organizational Science*, Thousand Oaks, CA: Sage Publications.

Sharkey, N. (2014) 'The Automatation and Proliferation of Military Drones and the Protection of Civilians', in Winston, M.E. and Edelbach, R. (eds), *Society, Ethics, and Technology*, Boston, MA: Wadsworth Cengage Learning, pp. 180–9.

Singer, P.W. (2009) *Wired for War: The Robotics Revolution and Conflict in the Twenty-First Century*, New York: Penguin Press.

Sparrow, R. (2009) 'Building a Better WarBot: Ethical Issues in the Design of Unmanned Systems for Military Applications', *Science and Engineering Ethics*, 15, 2: 169–87.

Stryber, R. (2001) 'Legitimacy', in Smelser, N.J. and Baltes, P.B. (eds), *International Encyclopedia of the Social and Behavioural Sciences*, Oxford/New York: Elsevier: 8700–4.

UN (United Nations High-level Panel on Threats, Challenges and Change) (2004) *A More Secure World: Our Shared Responsibility*, New York: United Nations Department of Public Information.

Vennesson, P. (2011) 'War Without the People', in Strachan, H. and Scheipers, S. (eds), *The Changing Character of War*, Oxford: Oxford University Press, pp. 241–58.

Whetham, D. (2013) 'Killer Drones: The Moral Ups and Downs', *The RUSI Journal*, 158, 3: 22–32.

Whetham, D. and Strawser, B.J. (2013) 'Eyes over Syria: Using Drones to Monitor Atrocities', *RUSI Analysis* (27 August 2013), www.rusi.org/analysis/commentary/ref:C521C9676C0B66 (accessed 12 January 2014).

Wiesner, I. (2013) *Importing the American Way of War? Network-Centric Warfare in the UK and Germany*, Militär und Sozialwissenschaften, Baden-Baden: Nomos.

Winner, L. (1986) 'Do artifacts have politics?', in *The Whale and the Reactor: A Search for Limits in an Age of High Technology*, Chicago: University of Chicago Press, pp. 19–39.

Part III

Implications for decision making at the strategic, operational, and tactical levels

9 Psychological perspectives on military decisions to deploy precision strike capabilities

Adrian Banks and Mandeep K. Dhami

Introduction

Decisions made by people, from policy makers through commanding officers to tactical operators, greatly influence what is achieved with precision strike capabilities. This chapter reviews different psychological perspectives on how these decisions are made and explores the difficulty in developing an effective process for making decisions about precision strikes. The review concludes that the complex and uncertain nature of such decisions suits a more intuitive strategy such as using simple heuristics, whereas the unfamiliar nature of these decisions requires analytic strategies. Therefore, these decisions are best made with a mixture of intuitive and analytic thought, a mode of thinking known as quasirationality. The exact mixture depends on features of the task, such as the amount of information available and degree of time pressure. This means that a single optimal decision process cannot be prescribed, instead a flexible combination of strategies are required in this domain.

The technological sophistication of today's precision strike capabilities, which include guided munitions ('smart bombs') and unmanned aerial vehicles (UAVs; 'drones') allow the military to achieve a higher level of attack precision than ever before, along with other benefits in surveillance. This has greatly influenced the increased use of these capabilities for tactical and strategic purposes. However, no matter how advanced the technology becomes, it can only be as effective as the humans who design, control, and use it. Decisions made by people, from policy makers through commanding officers to military operators, strongly determine what is achieved with precision strike capabilities. Although the use of smart bombs and drones has raised political, legal, and ethical concerns (e.g. Labot and Tawfeeq, 2013; O'Connell, 2010; Rogers, 2013; Strawser, 2010; Weizmann, 2013), to date there has been relatively little consideration of the issues pertaining to the 'human in the loop'. Understanding how decisions are made about precision strikes can inform debates as to their deployment as well as help decision makers at all levels of military operations appreciate the ultimately human face of such systems. In this chapter, we review different psychological perspectives on how humans may make decisions about precision strike capabilities.

These psychological perspectives lead us to ask two questions. First, how should 'rational' people make decisions about precision strike capabilities? From this normative standpoint, it is generally agreed that rational decision makers will collect, weight, and integrate all of the relevant information before making judgements and choices, and that they will choose on the course of action that is most likely to result in desired outcomes while also minimising any undesired consequences (e.g. Edwards, 1961; von Neumann and Morgenstern, 1947). But, as we will show, it is rarely the case that people (including professionals or experts) apply this approach.

The second question we therefore ask is: How do decision makers actually behave? From this descriptive standpoint we recognise that human cognitive abilities are limited and can be adversely affected by extraneous factors, and so people often rely on heuristic approaches to decision making (e.g. Gigerenzer and Goldstein, 1996; Tverksy and Kahneman, 1974). Heuristics are simple rules of thumb that require only a small amount of effort and information and yet generate a solution that is often good enough. Although heuristic decision making can often be both effective and cost-efficient (see Gigerenzer *et al.*, 1999), it may sometimes lead to biased and erroneous decisions (see Kahneman *et al.*, 1982). Before we discuss how the above two psychological perspectives apply to decision making about precision strike capabilities, we highlight the issues relevant to decision making in this domain.

The decision context

Precision strike capabilities bring many benefits. For example, they enable the operator to focus on numerous targets, in a quicker period of time, with a single weapon – and with greater precision. This makes military operations potentially faster, more effective, and efficient than when using other capabilities (Tiron, 2004). The enhanced precision reduces the chances of collateral damage. These capabilities can also be used in situations that would be too dangerous for manned aircraft, such as biologically or chemically contaminated environments. Finally, precision strike capabilities have the ability to protect areas that are outside the jurisdiction of any one nation (Huiss, 2012). These capabilities bring particular advantages that can influence the decision to use precision strikes.

Decisions about the use of precision strike capabilities are made at different levels in the chain of command, and different kinds of decisions are made along this chain. Policy makers decide if, in principle, precision strikes should be used, and this decision draws upon a wide range of factors such as political, legal, and ethical implications as well as military ones. An operational commander, on the other hand, must decide if a precision strike will be used in a given situation to achieve a specific objective, and so must consider the likelihood of achieving that objective. Finally, a tactical operator must decide whether to go through with a strike or not in a short time frame with little opportunity for reflection, and using the information available at that time.

Precision strike systems have fostered the notion in many countries that war is "blood-less and error-free" (Mahnken, 2011: 10). However this is naïve. Accuracy in precision strike capabilities is higher when specific targets are identified and located before an attack, using reliable and valid intelligence as well as reconnaissance missions. But if this is not possible then the identification and location occurs using intelligence gathered 'on the fly' during the battle and in the battle space. This can be less reliable and there is great opportunity for errors. Thus, the accuracy of decisions involving precision strike capabilities is heavily reliant on the reliability and validity of the information/intelligence gathered and used (Tiron, 2004). This means that, despite the precision of the technology, there is always the potential for errors to occur through inaccuracy in the decision process involved in deploying it.

The level of error and accuracy in decision making will be affected by factors internal to the decision maker as well as external to him/her, i.e. the decision context. Some of the extraneous factors that can potentially influence decision makers at different levels within the chain of command include stress and fatigue (Starcke and Brand, 2012), psychological distance (Mahnken, 2011), time pressure (e.g. Rieskamp and Hoffrage, 1999), information overload or insufficiency (e.g. Chewning and Harrell, 1990; Hwang and Lin, 1999; Jagacinski, 1994), how information is represented (e.g. Garcia-Retamero and Dhami, 2011, 2013), accountability (e.g. Lerner and Tetlock, 1999), system complexity (Huiss, 2012), legal rules (e.g. Weizmann, 2013), socio-political climate (e.g. Mintzberg, 1984), team factors (e.g. Banks and Millward, 2009), and organizational factors (e.g. Gore *et al.*, 2006). We will focus in particular on the difference between how rational decisions should be made and how decisions are actually made in the face of these challenges.

A distinction is often drawn between decisions made under risk and decisions made under uncertainty. A feature of decision making in this domain is that decisions are made under uncertainty. Decision making under risk occurs when the outcomes are not certain, but it is possible to estimate with some confidence the chances of events occurring in the future. In this case a balanced decision can be made based on numerical estimates of the likelihood of different scenarios. Decision making under uncertainty occurs when the likelihoods of future events are not well known and cannot easily be numerically estimated. In this case making a balanced decision is more difficult. Many decisions concerning precision strike capabilities are made under uncertainty because they rely on information that cannot be reliably quantified. We will explore different models of decision making under risk and uncertainty and the implications they have for the deployment of precision strikes.

How should rational people make decisions?

When making decisions about precision strike capabilities, decision makers must make both judgements (e.g. what is a good target?) and choices (e.g. which of these good targets should we strike?). Models of rational judgements and choices involve searching for and selecting the relevant information, weighting it optimally and then combining it to find the best solution. Those making decisions about precision strike capabilities might be expected to decide rationally to the extent that they should gather and appropriately use all of the relevant information before making a decision. Therefore it is important to understand models of rational decision making in order to compare them against human performance.

Rational models of judgement include statistical regression models and non-statistical weighted additive models. These propose that people search through all of the available information and weight and integrate it. For example, forming a judgement about whether a target is legitimate might involve a number of cues such as: presence of a known terrorist; target is an opposition leader; target has criminal/terrorist links outside the country; capture of target is difficult; removal of the target reduces opposition operations; and so on. To apply a weighted additive model to this situation, the value of each cue (e.g. known terrorist) is multiplied by the predetermined weight of that cue and these are summed across all cues. If the sum is greater than the threshold, then the target is judged to be legitimate. The emphasis in this model is the comprehensive use of information, the weighting of information so that each piece contributes an optimal proportion to the overall picture, and the compensatory integration of information so that a high value on one piece of information may compensate for an absence of threat indicators elsewhere (e.g. see Dhami, 2003; Dhami and Harries, 2001).

A similar weighted additive model can be applied when making a choice between two or more courses of action, such as where a strike should occur against a travelling target. In this case, each possible course of action is described in terms of its value on key attributes (e.g. route A may be clear from the air and go through an unpopulated area but has air defences, whereas route B is similar but does not have these air defences). For each alternative, the value of every attribute is multiplied by the predetermined weight of that attribute and these are summed. The course of action with the greatest sum is the optimal choice.

Finally, when making choices under risk and uncertainty, decision makers often need to consider the pros and cons of each alternative. The (subjective) expected utility model is a normative model depicting the idea that people weight and integrate the benefits and drawbacks of different alternatives before choosing amongst them (see Anand, 1995; Edwards, 1992; Fishburn, 1981). For example, a commander may decide whether to deploy a manned or unmanned aerial vehicle to strike a chosen target by assessing the importance of the possible benefits and drawbacks of each course of action (e.g. cost/resource/time-efficient, political/legal consequences, possible collateral damage, risk to personnel etc.), in addition to assessing the subjective probabilities of those possible outcomes occurring. The expected utility of the course of action should then be determined by the probability-weighted difference between the benefits and drawbacks. The more that the weighted difference favours the benefits, the more likely the decision maker ought to engage in that course of action.

How do people actually make decisions?

Peoples' ability to make judgements and choices relies on their basic cognitive processes, and these have adapted as the human brain has evolved. These abilities have neural bases, develop over the life-course, and can be trained and aided (see Dhami *et al.*, 2011). In recent years some psychologists have suggested that the decision models described above are not psychologically plausible (e.g. Dhami and Harries, 2001; Gigerenzer *et al.*, 1999). This is because they involve complex calculations involving a plethora of information. To do this accurately would require attention, memory, and computational capacities that are beyond the (unaided) human mind, and would be particularly difficult to achieve under certain conditions such as cognitive stress, time pressure, and limited information (e.g. Hammond, 2000; Payne *et al.*, 1993; Simon, 1990). In fact, there is ample empirical evidence to suggest that people do not engage in the sort of complex cognitive computations required by these models (see Gigerenzer *et al.*, 2011).

One solution to the problem of information overload proposed by Simon (1955) is to use heuristics. These are simple rules of thumb that require only a small amount of effort and information and yet generate a solution that is often good enough. A range of heuristics have been identified, some of which lead to good decisions and others that have the potential to generate biased decisions.

One set of simple heuristics has been proposed as an alternative model of how people make judgements and choices (Dhami, 2003; Dhami and Harries, 2001; Gigerenzer *et al.*, 1999). They are sometimes called 'fast and frugal' decision strategies because they rely on little information and make decisions rapidly. In addition, many of them are non-compensatory, meaning that once a decision is made on one piece of information it cannot be altered by other available information, no matter how relevant it may be.

An example is the Matching Heuristic for judgement tasks (Dhami, 2003; Dhami and Ayton, 2001). According to the Matching Heuristic, cues are searched in order of their utility. The individual looks for a critical value on each cue in turn. If a critical value is found, then a judgement can be made (e.g. this is a good target). If no critical value is found, then the individual searches for the next available cue. Search continues until the last available cue has been considered, and if that also has no critical value, then a default judgement is made (e.g. not a good target). Typically only a small number of cues are searched and only one piece is actually used (e.g. Cropp *et al.*, 2011) despite the wide range of information that rational models of decision making propose should be incorporated. Figure 9.1 provides an example of a hypothetical but typical matching heuristic to illustrate the simple nature of the decision processes proposed by this model of decision making. In this example, the same cues are available as in the example weighted additive model of judgement above. But whereas that model combined several cues using a relatively complex process, the model here includes only two cues and a simpler process whereby a decision is only ever made based on one cue.

The use of simple heuristics provides a significantly different perspective on how decisions about precision strikes are made if compared to the rational models proposed earlier. Given the major military, political, and legal ramifications of the decisions to use precision strike technology one might assume that the decision process would be complex and involved. However, it is in just these situations – where the decision is made under uncertainty – that simple heuristics can prove most effective (Gigerenzer and Goldstein, 1996). There is often too much information, too little time, and too much stress. A simple heuristic focuses only on the most important information necessary to make a satisfactory decision.

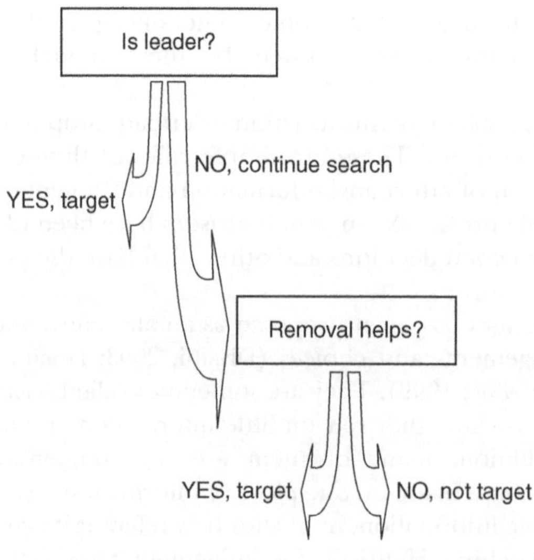

Figure 9.1 Matching heuristic for judging if this is a legitimate target.

Also, by focusing on the key information such heuristics are robust to changing circumstances that could prevent decision processes that rely on all available information from being readily implemented. If decisions are made about precision strikes using simple heuristics, it is important to appreciate which are the small number of cues that drive the decision rather than expecting all relevant information to be incorporated. It may well be that much apparently relevant information does not in practice influence the decision at all.

Not all heuristics are effective however. Kahneman and Tversky identified a number of heuristics which provide simple estimates of the likelihood of events that can be used when evaluating a course of action but which are prone to systematic biases that reduce the accuracy of decisions (Kahneman and Tversky, 1972, 1973, 1979). One well-known heuristic is the availability heuristic. This heuristic is used to assess the frequency or probability of an event by the ease with which instances or occurrences of it can be brought to mind. Often availability is a reasonable rule of thumb to use, but if the availability of an event is influenced by factors other than how frequently it has occurred in the past then this estimate can be erroneous. For example, the likelihood that a target is judged to be hostile is influenced by the ease with which previous experiences of similar targets that were hostile can be brought to mind. If, however, hostile targets have been encountered in the past that were very salient but also rare – for example, the cause of a major international incident – then these previous occurrences would be readily available in memory. Their salience would cause judgements of the likelihood that this particular target is hostile to be greater than it actually is.

There are many different heuristics that might be used in this domain and the consequences they have for the quality of the decisions varies. On the one hand, it is argued that simple heuristics can lead to effective decision making that has the advantage of requiring less information, time, and effort (see Gigerenzer *et al.*, 1999). Effective decisions are possible when heuristics rely on the cues or attributes that are the most valid predictors of a criterion variable (e.g. the most valid predictor of what is a legitimate target, e.g. see Snook *et al.*, 2011). On the other hand, in some circumstances the use of heuristics can lead to systematic and predictable biases. For example, the availability heuristic can cause a systematic overestimation of the likelihood of especially salient events. Either way, it is likely that decisions in this domain are influenced by simple rules of thumb that rely on small amounts of information (ignoring other information) to a greater extent than one might assume, given the apparently significant nature of the situation. Decisions can be improved not by attempting to limit the use of heuristics, but by understanding which heuristics are used and under what circumstances they lead to accurate or inaccurate decisions in this domain.

How do experts make decisions?

The decision models described above apply broadly across decision makers; however, some research has focused specifically on decisions made by experienced professionals to investigate how they develop and apply their expertise (e.g. Klein, 1998; Shanteau, 1988, 1992a, 1992b). This work is relevant in the current domain and so is worthy of discussion. Experts characteristically have very detailed knowledge about the situations they commonly face (e.g. Chase and Simon, 1973; Gobet and Waters, 2003). This knowledge is organised into templates or generic scenarios that frequently occur. When experts face a new situation they do not process many unconnected facts – that would be overwhelming. Instead, all of the facts fit into one of several familiar scenarios. Missing information can often be filled in using default values. The cognitive representations of experts are not simply more detailed than that of novices, they are qualitatively different (e.g. Chi *et al.*, 1981). Whereas novices classify problems according to their surface features, experts understand them in terms of more fundamental, deep structures. Experts also process information differently to novices. Garcia-Retamero and Dhami (2009) found that experts rely on fewer cues than novices in making a choice about properties that are vulnerable to burglars than novices and integrate this information differently to novices. The experience that experts have means that they make decisions that are not simply better informed than novices, they actually use different decision processes.

The decision making of experts is influenced by their domain knowledge. Klein has developed a recognition-primed decision making model of expert decision making in which most decisions involve little more than identifying the relevant scenario from their previous experiences (Klein, 1998). Through a process of pattern recognition, experts match the current situation to one of several typical scenarios that have been learnt over time. That scenario will have a default action to take. Therefore, rather than evaluating different options to find the best one, experts focus on understanding the situation, and from this they decide what to do simply because they have seen it before. For example, a number of indications from intelligence analysis may not individually be conclusive, but together they fit a pattern which indicates a potential threat that is worthy of further investigation, or potentially immediate action. This decision process is relevant in the context of precision strikes to so-called 'signature strikes' which are used against targets identified by a pattern of intelligence that indicates a significant operative or plot. The effectiveness of this decision process is influenced by the knowledge of the decision makers and the accuracy of the patterns that they recognise in the intelligence which are developed over considerable experience. However, expertise is only developed after lengthy, deliberate practice involving careful feedback (Ericsson *et al.*, 1993), and so when decisions are made but the full outcome is not known, as in some targeted killings, the experience of making the decision does not readily translate into improving the accuracy of the decision maker.

Quasirationality

The approaches to decision making that have been discussed so far can be broadly categorised as analytic and intuitive modes of thinking. Analytic thinking is thorough and rigorous: the type of thinking that is required when applying a rational decision making model. Intuitive thinking is fast and less effortful; the type of thinking that is required when applying a heuristic decision model or an expert recognising a familiar scenario. However, people do not exclusively think either analytically or intuitively. Both types of thought are important (e.g. Epstein, 1994; Evans and Over, 1996; Kahneman, 2003; Sloman, 1996).

Hammond (1996, 2000) developed this idea within Cognitive Continuum Theory (see Dhami and Thomson, 2012). Pure intuition and pure analysis fall at either end of the spectrum. In between lie 'quasirational' modes of thinking. These involve a combination of both intuitive and analytic thought. In practice, most decisions are quasirational although the amount of intuition and analysis used will vary. The theory suggests that decision making is most effective when the mode of cognition is well matched to the task. Intuitive modes of cognition are induced by various features of the tasks, including: tasks that are familiar; where there are many pieces of information to process; where the information is interpreted subjectively; where there are many possible responses; and when there is time pressure. Conversely, some features of tasks that induce analytic modes of thinking are: where the task is unfamiliar; where there are less than five pieces of information; where information can be interpreted objectively; where there are few response options; and when time pressure is low (Doherty and Kurz, 1996). Since cognitive continuum theory highlights the importance of the interaction between the human mind and the decision task, it is important to consider the features of the decision task to understand better how military decision makers make judgements and choices about precision strike capabilities.

Decisions about the use of precision strike capabilities are made at different levels in the chain of command, and different kinds of decisions are made along this chain. The features of the different tasks are likely to induce different modes of thinking. For instance, policy makers face ever changing circumstances and must respond to new and unanticipated events. Where these scenarios are novel and unfamiliar, analytic thought is encouraged because there is not a wealth of experience to draw upon. New problems are faced with no established solutions, for example the ethical and legal questions that are raised along with questions about how to make best tactical use of the new capabilities. These situations vary in time pressure, but often there will be time to reflect analytically before reaching a decision. On the other hand, other aspects of the task induce more intuitive thought. There is a range of information available and many possible options could be taken. This makes a complex, analytic decision process hard to follow and encourages intuition. Decisions at this level are well suited to a combination of analytic and intuitive thought.

Operational commanders must also face changing circumstances, but there may be more familiarity, i.e. similar missions in the past. The intelligence these commanders receive varies in reliability and cannot always be interpreted objectively, however. These features of the task are suited to more intuitive modes of cognition. On the other hand there is often time to consider decisions in depth, which supports analytic thinking. Decisions at this level are also well suited to a combination of analytic and intuitive thought.

Finally, decisions made by the tactical operator 'on the fly' are most suited to intuitive strategies. The time pressure is high as the target rapidly approaches, the situation is familiar from previous missions and training, and a range of information is presented simultaneously that must be processed at the same time. The individual may also experience cognitive stress. Decisions made by tactical commanders, like policy makers are better suited to a combination of intuitive and analytic thinking.

Decisions about precision strikes are therefore likely to be characterized by a mixture of intuitive and analytic thought – quasirationality. The exact mixture will depend on the situation that is being faced. This mixture means that decision making is not best supported by prescribing a single, formal, and systematic process for decision making. Such a process encourages analytic thought and so is useful in situations where this is well suited to the task, such as novel situations, but less effective in situations that are better suited to intuitive strategies, such as applying simple heuristics. A more flexible approach to decision making is required supporting both types of decision making, depending on the circumstances.

This suggests two key questions for further research. First, given that a range of decision strategies are required, what are those strategies? Simple heuristics are effective when they exploit features of the environment such as the validity of key cues. This means that heuristics are specific to certain decisions. Research eliciting the heuristics used by effective decision makers in each of these decision domains is required to discover which should be used and which should be avoided. Second, how might these decisions best be supported? Analytic decision processes can be supported using formal methods of evaluating criteria such as risks and outcomes. Heuristics can be supported using simpler checklists of key cues (e.g. Green and Mehr, 1997). Decision aids can be developed for both of these approaches through cognitive task analysis (e.g. Crandall *et al.*, 2006). However, as quasirationality requires a mixture of approaches, a further question is when to move between the decision methods to ensure that they are best suited to the decision being made.

Summary and conclusions

Rational models of decision making involve a complex process of selecting, weighting, and combining relevant information. However, the actual process of making a decision often does not often follow this analytic model. Given the significance of these situations, one might assume that the decision process would involve a similarly complex weighting of a wide range of information. But the human mind has a limited capacity and cannot readily process all of this information. Instead, it is likely that decisions will be influenced by heuristics that use only a small amount of the information available. Some heuristics are effective and others lead to systematic biases, and so it is important to discover which information is influential in the decision and which can be ignored, as well as reduce any opportunity for bias. Experts also often use less information, but they draw upon their previous experiences to focus on the key information and to interpret this information and place it within familiar scenarios that they use to make decisions.

Decisions vary from detailed rational analyses of the situation to more intuitive processes based on simple heuristics and limited information. The nature of the task determines where on this spectrum is most effective for making decisions. This preliminary analysis suggests that the complex and uncertain nature of the decisions surrounding precision strikes suits a more intuitive strategy, whereas the unfamiliar nature of the decisions requires analysis to generate novel solutions. Therefore, these decisions are best made with a mixture of intuitive and analytic thought, a mode of thinking known as quasirationality. Supporting quasirational models of thinking means that a single, formal decision process cannot be described to govern decisions. A more flexible approach is required to support the optimal mixture of different decision strategies that are best suited to the situation.

References

Anand, P. (1995) *Foundations of Rational Choice under Risk*, Oxford: Oxford University Press.

Banks, A.P. and Millward, L.J. (2009) 'Distributed Mental Models: Mental Models in Distributed Cognitive Systems', *The Journal of Mind and Behavior*, 30: 249–66.

Bröder, A. (2000) 'Assessing the Empirical Validity of the "Take-The-Best" Heuristic as a Model of Human Probabilistic Inference', *Journal of Experimental Psychology: Learning, Memory, and Cognition*, 26: 1332–46.

Campitelli, G. and Fernand, G. (2011) 'Deliberate Practice: Necessary but not Sufficient', *Current Directions in Psychological Science*, 20: 280–5.

Chase, W.G. and Simon, H.A. (1973) 'Perception in Chess', *Cognitive Psychology*, 4: 55–81.

Chewning, E.G. and Harrell, A.M. (1990) 'The Effect of Information Load on Decision Makers' Cue Utilization Levels and Decision Quality in a Financial Distress Decision Task', *Accounting, Organizations and Society*, 15: 527–42.

Chi, M.T., Feltovich, P.J., and Glaser, R. (1981) 'Categorization and Representation of Physics Problems by Experts and Novices', *Cognitive Science*, 5: 121–52.

Crandall, B., Klein, G.A., and Hoffman, R.R. (2006) *Working Minds: A Practitioner's Guide to Cognitive Task Analysis*, Cambridge, MA: MIT Press.

Cropp, N., Banks, A., and Elghali, L. (2011) 'Expert Decision Making in a Complex Engineering Environment: A Comparison of the Lens Model, Explanatory Coherence, and Matching Heuristics', *Journal of Cognitive Engineering and Decision Making*, 5: 255–76.

Dhami, M.K. (2003) 'Psychological Models of Professional Decision-Making', *Psychological Science*, 14: 175–80.

Dhami, M.K. and Ayton, P. (2001) 'Bailing and Jailing the Fast and Frugal Way', *Journal of Behavioral Decision Making*, 14: 141–68.

Dhami, M.K. and Harries, C. (2001) 'Fast and Frugal versus Regression Models of Human Judgment', *Thinking and Reasoning*, 7: 5–27.

Dhami, M.K. and Harries, C. (2010) 'Information Search in Heuristic Decision Making', *Applied Cognitive Psychology*, 24: 571–86.

Dhami, M.K. and Mandel, D.R. (2012a) 'Crime as Risk Taking', *Psychology, Crime and Law*, 18: 389–403.

Dhami, M.K. and Mandel, D.R. (2012b) 'Forecasted Risk Taking in Youth: Evidence for a Bounded-Rationality Perspective', *Synthèse*, DOI: 10.1007/s11229-012-0110-2.

Dhami, M.K. and Mandel, D.R. (2013) 'How Do Defendants Choose their Trial Court? Evidence for a Heuristic Processing Account', *Judgment and Decision Making*, 8: 552–60.

Dhami, M.K. and Thomson, M. (2012) 'On the Relevance of Cognitive Continuum Theory for Understanding Management Judgment and Decision Making', *European Management Journal*, 30: 316–26.

Dhami, M.K., Schlottmann, A., and Waldmann, M. (eds) (2011) *Judgment and Decision Making as a Skill: Learning, Development, and Evolution*, Cambridge: Cambridge University Press.

Doherty, M.E. and Kurz, E.M. (1996) 'Social Judgement Theory', *Thinking and Reasoning*, 2: 109–40.

Edwards, W. (1961) 'Behavioral Decision Theory', *Annual Review of Psychology*, 12: 473–98.

Edwards, W. (ed.) (1992) *Utility Theories: Measurements and Applications*, Boston, MA: Kluwer.

Epstein, S. (1994) 'Integration of the Cognitive and the Psychodynamic Unconscious', *American Psychologist*, 49: 709–24.

Ericsson, K.A., Krampe, R.T., and Tesch-Römer, C. (1993) 'The Role of Deliberate Practice in the Acquisition of Expert Performance', *Psychological Review*, 100: 363–406.

Evans, J.B. and Over, D. (1996) *Rationality and Reasoning*, Hove, UK: Psychology Press.

Fishburn, P.C. (1981) 'Subjective Expected Utility: A Review of Normative Theories', *Theory and Decision*, 13: 139–99.

Franke, U.E. (2013) 'The Five Most Common Media Misrepresentations of UAVs', in Aaronson, M. and Johnson, A. (eds), *Hitting the Target? How New Capabilities are Shaping International Intervention*, London: Royal United Services Institute, pp. 19–31.

Garcia-Retamero, R. and Dhami, M.K. (2009) 'Take-The-Best in Expert-Novice Decision Strategies for Residential Burglary', *Psychonomic Bulletin and Review*, 16: 163–9.

Garcia-Retamero, R. and Dhami, M.K. (2011) 'Pictures Speak Louder than Numbers: On Communicating Medical Risks to Immigrants with Non-Native Language Proficiency', *Health Expectations*, 14 (Supp. 1): 46–57.

Garcia-Retamero, R. and Dhami, M.K. (2013) 'On Avoiding Framing Effects in Experienced Decision Makers', *Quarterly Journal of Experimental Psychology*, 66: 829–42.

Gigerenzer, G. and Goldstein, D. (1996) 'Reasoning the Fast and Frugal Way: Models of Bounded Rationality', *Psychological Review*, 103: 650–69.

Gigerenzer, G., Hertwig, R., and Pachur, T. (eds), (2011) *Heuristics: The Foundations of Adaptive Behaviour*, New York: Oxford University Press.

Gigerenzer, G., Todd, P.M., and ABC Research Group (1999) *Simple Heuristics that Make us Smart*, Oxford: Oxford University Press.

Gilovich, T., Griffin, D., and Kahneman, D. (2002) *Heuristics and Biases: The Psychology of Intuitive Judgment*, Cambridge: Cambridge University Press.

Gobet, F. and Waters, A.J. (2003) 'The Role of Constraints in Expert Memory', *Journal of Experimental Psychology: Learning, Memory, and Cognition*, 29: 1082–94.

Gore, J., Banks, A., Millward, L.J., and Kyriakidou, O. (2006) 'Naturalistic Decision Making and Organisations: Reviewing Pragmatic Science', *Organization Studies*, 27: 925–42.

Green, L., and Mehr, D.R. (1997) 'What Alters Physicians' Decisions to Admit to the Coronary Care Unit?', *Journal of Family Practice*, 45: 219–26.

Hammond, K.R. (1996) *Human Judgment and Social Policy: Irreducible Uncertainty, Inevitable Error, Unavoidable Injustice*, Oxford: Oxford University Press.

Hammond, K.R. (2000) *Judgments under Stress*, New York: Oxford University Press.

Huiss, R. (2012) *Proliferation of Precision Strike. Issues for Congress*, www.fas.org/sgp/crs/nuke/R42539.pdf (accessed 12 January 2014).

Hwang, M.I. and Lin, J.W. (1999) 'Information Dimension, Information Overload and Decision Quality', *Journal of Information Science*, 25: 213–18.

Jagacinski, C.M. (1994) 'Evaluation of Job Candidates with Missing Information: Effects of Attribute Importance and Interattribute Correlation', *Journal of Behavioral Decision Making*, 7: 25–42.

Kahneman, D. (2003) 'A Perspective on Judgment and Choice: Mapping Bounded Rationality', *American Psychologist*, 58: 697–720.

Kahneman, D. and Tversky, A. (1972) 'Subjective Probability: A Judgment of Representativeness', *Cognitive Psychology*, 3: 430–54.

Kahneman, D. and Tversky, A. (1973) 'On the Psychology of Prediction', *Psychological Review*, 80: 237–51.

Kahneman, D. and Tversky, A. (1979) 'Prospect Theory: An Analysis of Decision under Risk', *Econometrica*, 47: 263–91.

Kahneman, D., Slovic, P., and Tversky, A. (eds) (1982) *Judgment under Uncertainty: Heuristics and Biases*, Cambridge: Cambridge University Press.

Klein, G. (1998) *Sources of Power: How People Make Decisions*, Cambridge, MA: MIT Press.

Labot, E. and Tawfeeq, M. (2013) *Drone Strikes Kill Militants in Yemen. Americans Urged to Leave*, CNN, http://edition.cnn.com/2013/08/05/world/yemen-us-drone-strike/?hpt=hp_t2 (accessed 12 January 2014).

Lerner, J.S. and Tetlock, P.E. (1999) 'Accounting for the Effects of Accountability', *Psychological Bulletin*, 125: 255–75.

Mahnken, T.G. (2011) *Weapons: The Growth and Spread of the Precision-Strike Regime*, http://legacy2.sais-jhu.edu/pressroom/pdf/2011_06_01_mahnken.pdf (accessed 12 January 2014).

Mintzberg, H. (1984) 'Power and Organizational Life Cycles', *Academy of Management Review*, 9: 207–24.

O'Connell, M.E. (2010) 'Unlawful Killing with Combat Drones: The Case of Pakistan 2004–2009', in Bronitt, S. (ed.), *Shooting to Kill: The Law Governing Lethal Force in Context*, Notre Dame Legal Studies Paper 09–43, http://papers.ssrn.com/sol3/papers.cfm?abstract_id=1501144 (accessed 12 January 2014).

Pachur, T. and Marinello, G. (2013) 'Expert Intuitions: How to Model the Decision Strategies of Airport Customs Officers?', *Acta Psychologica*, 144: 197–103.

Payne, J.W., Bettman, J.R., and Johnson, E. (1993) *The Adaptive Decision Maker*, Cambridge: Cambridge University Press.

Rieskamp, J. and Hoffrage, U. (1999) 'When Do People Use Simple Heuristics, and How Can We Tell?', in Gigerenzer, G., Todd, P.M., and the ABC Research Group (eds), *Simple Heuristics That Make Us Smart*, New York: Oxford University Press, pp. 141–67.

Rogers, J.F. (2013) 'The Public View: British Attitudes to Drone Warfare and Targeted Killing', in Aaronson, M. and Johnson, A. (eds), *Hitting the Target? How New Capabilities are Shaping International Intervention*, London: Royal United Services Institute, pp. 7–17.

Shanteau, J. (1988) 'Psychological Characteristics and Strategies of Expert Decision Makers', *Acta Psychologica*, 68: 203–15.

Shanteau, J. (1992a) 'Competence in Experts: The Role of Task Characteristics', *Organizational Behavior and Human Decision Processes*, 53: 252–66.

Shanteau, J. (1992b) 'How Much Information Does an Expert Use? Is it Relevant?', *Acta Psychologica*, 81: 75–86.

Simon, H.A. (1955) 'A Behavioral Model of Rational Choice', *Quarterly Journal of Economics*, 69, 99–118.

Simon, H.A. (1987) *Models of Man*, New York: Garland.

Simon, H.A. (1990) 'Invariants of Human Behavior', *Annual Review of Psychology*, 41: 1–19.

Sloman, S. (1996) 'The Empirical Case for Two Systems of Reasoning', *Psychological Bulletin*, 119: 3–22.

Snook, B., Dhami, M.K., and Kavanagh, J. (2011) 'Simply Criminal: Predicting Burglars' Occupancy Decisions with a Simple Heuristic', *Law and Human Behavior*, 35: 316–26.

Stanovich, K.E. and West, R.F. (2000) 'Individual Differences in Reasoning: Implications for the Rationality Debate?', *Behavioral and Brain Sciences*, 23: 645–726.

Starcke, K. and Brand, M. (2012) 'Decision Making under Stress: A Selective Review', *Neuroscience and Biobehavioral Reviews*, 36: 1228–48.

Strawser, B.J. (2010) 'Moral Predators: The Duty to Employ Uninhabited Aerial Vehicles', *Journal of Military Ethics*, 9: 342–68.

Tiron, R. (2004) *US Army Assesses Precision Strike Capabilities*, www.nationaldefensemagazine.org/archive/2004/March/Pages/US_%20Army%20_Assesses3631.aspx (accessed 12 January 2014).

Tversky, A. and Kahneman, D. (1974) 'Judgment under Uncertainty: Heuristics and Biases', *Science*, 185: 1124–31.

von Neumann, J. and Morgenstern, O. (1947) *The Theory of Games and Economic Behaviour*, New York: Princeton University Press.

Voss, J.F., Greene, T.R., Post, T.A., and Penner, B.C. (1983) 'Problem-Solving Skill in the Social Sciences', *The Psychology of Learning and Motivation*, 17: 165–213.

Weizmann, N. (2013) 'Remotely Piloted Aircraft and International Law', in Aaronson, M. and Johnson, A. (eds), *Hitting The Target? How New Capabilities are Shaping International Intervention*, London: Royal United Services Institute, pp. 33–44.

10 Unmanned weapons systems and just wars

The psychological dimension

Nicola Power, Laurence Alison, and Jason Ralph

Introduction

This chapter explores the psychological impact of unmanned weapons systems on military decision making. Just war theory is used as a normative framework to evaluate whether decision making is morally acceptable before, during, and after conflict. We discuss the psychology of decision making at the strategic, operational, and tactical levels, and question whether the novelty of unmanned weapons systems can bias the consideration of ethics during conflict. We provide theoretical hypotheses that describe how the possession of unmanned weapons systems may influence military decisions. We query whether military decision makers have an appropriate level of understanding and expertise in using these systems; importantly when considering the trade-off between short-term tactical advantage and long-term strategic goals, and offer recommendations for research.

Defining ethics and technology

Prior to a more definitive analysis of the psychology of decision making, it is important to outline the theoretical foundation upon which this chapter is based. By their very nature wars and violent conflict equate to death and destruction, yet society tries to regulate the level of violence through consideration of internationally regarded ethical, moral, and legal normative standards. Just war theory (JWT: also known as the 'laws of war', Evans, 2005) is a centuries-old normative framework of moral standards for how decisions about conflict *ought* to be taken before, during, and after war. These standards are a 'heuristic tool, providing the set of moral criteria that should inform decisions on whether to go to war and, if so, how it should be fought' (Evans, 2005: 11); and their principles have been woven into various international treaties of law (e.g. Hague Conventions 1907; Geneva Conventions 1949). For example the *ad bellum* principle of 'last resort' is enshrined in Article 1 of the Hague Convention (III; 1907) by requiring 'reasoned declaration of war or an ultimatum'; and the UK

Ministry of Defence outlines *in bello* principles in the 'Joint Service Manual of the Law of Armed Conflict' (2004) as 'a violation of the law of armed conflict by the armed forces of a state involves the international responsibility of that state' (Joint Service Publication 383, 2004: 4). The concept of *jus post bellum* is more recent (Orend, 2000) but provides a useful addition to JWT in the context of contemporary international intervention. It is not the intention of the authors to provide a detailed critique of JWT. But for the purposes of this chapter we take it to define a set of moral standards that accept that, although conflict between states is inevitable, global society should try at least to keep conflict as humane, proportionate, and just as possible.

With this consideration of 'just wars' in mind, this chapter provides a psychological discussion of military decision making and the use of contemporary unmanned weapons systems (UWS). These are novel and revolutionary as they provide the military capability to send technology into conflict independent of direct human contact. They differ from other means of long-range force projection because of their ability to combine sustained over-watch with immediate destructive effects. UWS are used for a variety of tasks, from third-party targeting to reconnaissance and intelligence gathering (Kurkcu *et al.*, 2012). Examples of such systems include the MQ1-Predator drone equipped with Hellfire missiles and MQ-9 Reaper with laser-guided bombs.

We ask whether there has been sufficient time for military decision makers to develop expertise and fully understand the short, medium, and long-term effects of UWS on the battlefield. This chapter's argument is theoretical in scope and, by setting up opposing hypotheses against each other, intentionally provocative in nature. It is based upon two grounding principles. First, actions taken during conflict should be, as far as possible, just. This is something that, as outlined above, is reflected through various international legal treaties on how conflict ought to be conducted. Second, UWS are a relatively novel technology. Thus, although UWS can be highly advantageous for enabling effective short-term precision strikes, from a psychological dimension we query whether military decision makers yet fully understand the short, medium, and long-term consequences of their use. Fundamentally, a war should be fought as morally as possible, but does our relative inexperience of using UWS impede the ability to make objective 'just' decisions during conflict? The proliferation of such systems is occurring exponentially; A recent report for the US Congress stated that they 'are expected to take on every type of mission currently flown by manned aircraft' (Gertler, 2012: 6) and the global investment in unmanned technology is on the rise with a predicted compound annual growth rate of 4.4 per cent between 2012 and 2022 across North America, Europe, Asia, Middle East, Latin America, and Africa (Strategic Defence Intelligence, January 2013). Therefore, as with any modern technology it is important to ensure that this does not occur in the absence of objective and reasoned human comprehension of the limits to their use.

Theoretical hypotheses for how the use of unmanned weapons systems may influence the 'just' consideration of decisions on conflict

We will now outline three pairs of theoretical and opposing hypotheses (Table 10.1) that we have established following a psychological consideration of decision making: two for each stage of war (i.e. *ad bellum, in bello, post bellum*). The hypotheses are coupled during each stage of war as counter-hypotheses, i.e. one hypothesis suggests UWS can make war *more* just, whereas the other suggests UWS make war *less* just. Before exploring these in more detail however, we will first provide some background on the psychology of decision making.

Psychology of decision making

A decision is the commitment to a course of action that is taken in order to achieve a desired goal (Yates, 2003). The process of making a decision involves choosing between potential courses of action (i.e. options) within an uncertain environment (Hastie, 2001). Decision makers must first establish their situation awareness, formulate, recognise, and trade-off potential plans of action, and then finally execute their choice (van den Heuvel, Alison, and Crego, 2012). In an ideal environment, decision makers use pure rationality (i.e. expected utility theory) to analyse objectively and select the optimal option available (von Neumann and Morgenstern, 1947). Yet real-world decision making is bounded by environmental pressures, such as time pressure and ambiguity (Kruglanski and Thompson, 1999; Souchon *et al.*, 2009), which can lead to cognitive overload and poor decision making (Renkl *et al.*, 2009). Pressures within the decision making environment make option processing and trade-off more difficult and so decision makers will attempt to avoid or delay their choice (Anderson, 2003; van den Heuvel, Alison, and Crego, 2012) or rely on 'cognitive shortcuts' and 'heuristics' that can bias decision outcomes (DiBonaventura and Chapman, 2008; Tversky and Kahneman, 1974). For example, both the 'representative' heuristic (assume high commonality between prototypical entities) and 'confirmation bias' (search for cues in the environment which confirm one's assumption) can make the decision maker subjectively in favour of a predetermined option (Kellogg, 1995; Lord *et al.*, 1979). Decision making in uncertain environments must therefore be timely but not degraded by biased processing.

When a person is experienced in a decision domain they are able to make efficient and intuitive decisions even whilst placed under extreme pressure (Klein, 1998). The dual-process model describes how individuals

Table 10.1 Summary of theory-driven hypotheses on how UWS may influence the 'just' consideration of conflict before, during, and after war

Influence of UWS on decision making	*Reason why*
Ad bellum	
The decision to enter conflict becomes *more likely*	• UWS are perceived as a dominant and winning technology, especially when coupled with loss aversive motivations • There is increased public support for conflict due to perceptions of reduced ethnocentric risk of harm
The decision to enter conflict becomes *less or no more likely*	• Asymmetric dominance increases preference for diplomatic discussion • UWS are not perceived as overwhelmingly dominant and so have little or no effect on decision making
In bello	
Use of force during conflict becomes *more extreme*	• There is reduced ability to discriminate between targets (poor sensory information; increased complexity; biased situation recall) • There is uncertainty over whether they are proportional or necessary weapons as their short and long-term impact is unknown • Removing lethal force decisions from the frontline can induce moral disengagement and more extreme violence
Use of force during conflict becomes *less extreme*	• Moral disengagement can foster more analytical and ethical decision making • The next generation of more autonomous systems can selectively and accurately discriminate between targets
Post bellum	
UWS are counterintuitive to long-term strategic aims and perpetuate conflict	• They reduce the number of soldiers on the frontline which limits opportunity to establish trust and cohesion with intervened upon communities • They can increase anti-intervention sentiment and desire for revenge in intervened upon communities • They can increase the prevalence of radicalisation and extremist support in intervened upon communities
UWS facilitate long-term strategic aims and facilitate peace	• They are a necessary and 'least-worst' option to eradicate extremists within targeted regions

make sense of information via both intuitive (innate, non-taxing) and analytic (effortful, systematic) processing. When an individual is experienced in the decision domain, then they possess accurate intuitive skills known as domain-specific expertise (Kahneman and Frederick, 2002; Kahneman and Klein, 2009). Theories on metacognition (i.e. thinking about one's thinking) distinguish between three forms of expertise (Schraw and Moshman, 2005). At a basic level a novice possesses 'declarative knowledge' about *what* the decision domain is (e.g. 'I know what an UWS is'). At the middle level, a competent decision maker holds 'procedural knowledge' on *how* to execute choice (e.g. 'I know how to use an UWS'). And finally, at an expert level, the decision maker has 'conditional knowledge' about both *when* and *why* an action is appropriate (e.g. 'I know when and why I should fly an UWS'). In other words, an expert can make rapid decisions under extreme pressure, yet is able to maintain a rational and considered understanding of why and when certain courses of action are appropriate. Decision making improves as metacognitive knowledge increases.

Metacognition is the fundamental goal of all learning and the cornerstone of expertise (White and Frederiksen, 2005). It is needed for effective decision making as it allows experts to plan, monitor, and evaluate their choice, despite operating under high environmental and cognitive pressure (Akturk and Sahin, 2011). In order to achieve metacognitive expertise, a learner must plan, monitor, and evaluate their learning and actions by practicing in the domain-specific environment (Schraw and Moshman, 2005). When choice environments are uncertain, time pressured, and cognitively overloading, it is important that decision makers have the opportunity to develop expertise to be effective. This chapter assumes that the relative novelty of UWS means that there is a reduced level of expertise in their application. Although new technologies have the potential to improve decision making, there is a risk that decision makers do not fully understand the short, medium, and long-term consequences of their use. This leads us to consider whether using UWS may make decisions on conflict more or less morally just.

Ad bellum *(before war)*

According to JWT, conflict should only occur if there is just cause (i.e. self-defence or counter-action to punish severe actions that endangered innocent life), right intention (i.e. must only act in order to achieve this cause), and as a last resort (i.e. all peaceful options have been exhausted and failed) (Orend, 2008). Political decision makers must have a reasonable and just rationale for why they enter conflict with another state. Our first two theoretical counter-hypotheses consider how the availability of UWS as a potential option in the decision making process may influence choice outcome. Namely, whether the decision to enter conflict becomes *more* likely when UWS are an option; or whether the decision to enter conflict becomes *less or no more* likely when UWS are an option?

When individuals consider option-based decisions (e.g. whether to enter conflict or not) then they will try to trade off their options to select the one with the most attractive anticipated outcome(s) (Kahneman, 2003). They attempt to both maximise the potential gains from their choice, whilst also minimising losses (Kahneman and Tversky, 1979). In fact decision makers take considerably *more* effort in trying to minimise anticipated losses than they do to maximise gains: a psychological phenomenon known as loss aversion (Kahneman, 2003). The salient negative emotional impact of losing feels stronger than the positive emotional impact of an equal win; which makes individuals more risk taking when they anticipate potential for loss (Tversky and Kahneman, 1992). When considering 'do or don't act' decisions whereby one is deliberating over whether to take a course of action or not (e.g. decision to enter conflict or not?), decision makers are further more likely to avoid action; a psychological phenomenon known as 'omission bias' (Baron and Ritov, 1994). The combined effect of these two phenomena is that individuals try to avoid taking action (i.e. omission bias) and this desire to avoid action can be increased by any associated and emotionally salient potential for loss (i.e. loss aversion). The exception is when potential loss is anchored to *not acting* which can make decision makers more likely to take action.

In order to provide a worked example for how psychological phenomena such as loss aversion and omission bias may influence military decision making, we will discuss the 2003 invasion of Iraq. Despite the uncertainty and limited intelligence surrounding Iraq's potential possession of weapons of mass destruction, a decision was made to enter conflict. Potential loss was associated to *not acting* and thus decision makers were (arguably) biased to favour conflict. Chan (2012) argued that loss aversion motivates military choice as leaders prefer to 'gamble for recovery' than accept (potential) defeat or harm. The rationale for entering conflict could therefore have been motivated by loss aversion anchored to potential harm from inaction. In addition, decision makers also consider anticipated positive outcomes of their choice and thus these must be considered in tandem. Indeed spreading democracy was seen as a potential positive outcome and the possession of UWS may have made action seem less risky due to perceived military dominance. UWS make war seem less dangerous to the public as combatants from home nations are not put in direct danger. Indeed, public opinion and emotional reactions are one of the largest influences on political decision making (Muller, 2011) especially when facing a foreign policy crisis (Knecht and Weatherford, 2006). Singer (2008) has warned how removing the soldier from the frontline may make war more palatable to the public due to the reduced risk of allied loss of life. There is thus a danger that UWS may make war seem less dangerous to the public. From an *ad bellum* just war perspective, conflict is therefore more likely due to salient loss aversive motivations that are exaggerated by anticipated ethnocentric gains from utilising UWS.

An alternative view is that the possession of UWS has no effect on the likelihood of conflict or may even reduce its likelihood by encouraging diplomatic debate. Arkin (2009) has suggested that the geopolitical asymmetry in the possession of new technologies may indirectly reduce the chances of future conflict. He suggests that those states or actors who lack UWS will proactively favour diplomacy rather than risk drone attacks. Furthermore, the ethnocentric advantages outlined above may not be all that impactful. Indeed, the salient impact of potential loss is weighted more heavily than gains and so public perceptions may have little influence. Specifically, the advantages in improving the accuracy of complex attacks (Schulzke, 2011; Sharkey, 2008), and for overcoming human errors associated with fear, fatigue, and resistance to killing (Daddis, 2004; Grossman, 1995; Olsen *et al.*, 2010) does not mean that allied combatants are free from harm. Currently no war can be fought from complete remoteness and thus rudimentary weapons (such as improvised explosive devices) can cause significant casualties to those who remain on the frontline (Human Rights Watch, 2008). Technological innovation is also available to a wide range of actors in modern conflict, as shown by the use of smart phones in the Mumbai bombings (Goodman, 2011) and Hezbollah's claimed ownership of a rudimentary drone (BBC, 11 October 2012). As such, the motivation for entering conflict and biasing (or disregarding) just war considerations due to the possession of UWS is open to debate.

Thus our two counter-hypotheses are as follows. First, UWS can make war more likely as they act to strengthen loss aversive motivations via anticipated positive outcomes associated with military dominance and public support. Yet equally, UWS can deter *ad bellum* decisions to enter conflict by making diplomacy more favourable in the eyes of the opposition and the persistent risks still associated with conflict. Both of these hypotheses are counter-intuitive and purely theoretical. As UWS are a relatively new capability, with modern systems such as Predator and Reaper drones only having played a predominant role in international conflict since the turn of the century (Singer, 2009), experience with how UWS influence decision making is limited. *Ad bellum* decisions are motivated by risk assessment, trade-off, and cost benefit analysis, which can be psychologically influenced by anticipated positive and negative consequences (Kahneman, 2003). Cumulatively, cognitive biases can bias the rational consideration of JWT; something that is currently poorly measured due to their relatively recent advancement and thus needs empirical attention.

In bello *(during war)*

Actions taken during conflict ought to be just (Evans, 2005). The *in bello* principle of discrimination requires that lethal force effectively discriminates between legitimate targets and innocent civilians. The application of JWT during war therefore aims to moderate how extreme the acts of violence taken during conflict are. Our two opposing hypotheses *in bello* are that: (i) use of force during conflict is *more* extreme when possessing UWS; and conversely that (ii) use of force during conflict is *less* extreme when possessing UWS.

Critics of UWS argue that the relative deprivation of sensory cues available to the remote operator mean that decisions taken during conflict cannot be objective or discriminative (Sharkey, 2008; Sparrow, 2007). UWS cockpits, in comparison to traditional manned aircraft, lack the peripheral visual, haptic, and auditory cues usually experienced in live flight (Kurkcu *et al.*, 2012). The limitation of sensory cues can degrade the ability of the operator to conduct discriminative lethal force by increasing uncertainty and complexity; when a task is complex, then the ability to process sensory cues decreases (Sherman *et al.*, 2003). The separation of pilots from the usual sensory cues found in standard aircraft has been highlighted as an unprecedented psychological factor that can influence decision making (Reardon, 2013). As described earlier, domain-specific expertise can help individuals cope with complexity and make efficient decisions (Kahneman, 2003). Yet UWS are relatively novel and developing exponentially and thus it is important to question whether their operators yet possess sufficient expertise to cope with complexity.

The negative impact of complexity on decision tasks is especially of note when decisions require ethical consideration due to the potential loss of human life. It is important that decision makers hold a metacognitive awareness of the implications of their actions. They must understand not only how to use UWS, but under what conditions and boundaries their use is appropriate. Of course it is acknowledged that a decision making team is involved in these choices, but the operators themselves must also possess an awareness of rationale for their own moral accountability. Metacognition makes decision making more effective and importantly provides individuals with a salient rationale that can justify their actions. This is something that would no doubt be of benefit to the UWS operators who, like manned aircraft pilots, have been found to be at risk of mental health problems following combat (Otto and Webber, 2013). Complex environments impede an individual's ability to recall events accurately; causing them to generate inaccurate and stereotypical-consistent memories (Sherman *et al.*, 2003). Domain-specific metacognitive expertise would help to alleviate this effect.

Royakkers and van Est (2010) argue that the use of automation during conflict can also induce moral disengagement in the 'human operator' of such systems, which can have problems for discrimination. Moral disengagement, they argue, increases 'remote violence' due to the physical and psychological distance between the operator and the battlefield and can induce dehumanisation and detachment in lethal force decisions

(Bandura, 1986; Detert *et al.*, 2008). Although moral disengagement may be tactically useful for reducing resistance to killing in soldiers, it can impede battlefield effectiveness and cause negative psychological implications after returning to civilian life (Holmes, 2003; Lindlaw, 2008). It also raises ethical concerns over the soldier's awareness of the consequences of their actions: the death of another human being (Sparrow, 2009). Furthermore, it is only more recently that the mental health of UWS operators has created debate and discussion (Otto and Webber, 2013; Reardon, 2013) indicating a paucity of past consideration for their psychological effects. UWS operators are thus at risk to a triple-pressure effect on their ability to make ethical discriminative decisions *in bello* due to the deprivation of sensory cues, increased cognitive complexity, and impeded rationale and recall.

UWS may also impede the ability to judge whether actions taken during conflict are 'proportionate' (i.e. whether the potential for collateral damage is proportional to potential military gain) and 'necessary' (i.e. the minimum force necessary to achieve aims) (Orend, 2008). Singer (2008) has suggested that UWS create an overwhelming and unfair asymmetric advantage in favour of those who possess them; thus their use is neither proportionate nor necessary. There is currently very little information on the sheer magnitude of suffering or collateral cost caused by these systems and thus this is a contentious issue with limited data upon which to judge it (Asaro, 2006). Put simply, UWS have not been around long enough for there to an objective understanding of their implications; both in terms of the proportionality of their potential collateral damage (e.g. loss of property and/or life) and whether they are 'necessary' tools to achieve strategic military aims. Indeed, UWS may be counterintuitive in the long-term if the use of these systems creates opposition from innocent civilians who are 'intervened upon' in the conflict zone (e.g. civilians in Afghanistan).

From an alternative perspective, UWS may *increase* ethical considerations of decisions taken during conflict. Moral disengagement could in fact be useful for overcoming human errors associated with decision inertia (the failure to execute a choice due to redundant deliberation). When a decision maker feels accountable and uncertain about the fallout of their decisions in ambiguous and high stakes environments, they will try and avoid making a decision (Anderson, 2003). Rather than commit to a choice, they fail to implement action (van den Heuvel, Alison, and Crego, 2012). This is due to the negative affect and anticipated regret associated with making the wrong choice (Inman and Zeelenberg, 2002), and may manifest behaviourally as *omission* (when the decision is ignored) (Anderson, 2003), *choice deferral* (when decisions are delayed until further information is acquired) (Kopylov, 2009), *inaction inertia* (when the decision is delayed because of a missed opportunity) (O'Keefe and Wright, 2010), or *implementation failure* (when a decision may be made but no action taken) (van den Heuvel, Alison, and Crego, 2012). Ultimately, the experience and anticipation of uncertainty and negative emotions can either temporarily or permanently prevent decisive action (van den Heuvel, Alison, and Power, 2012). Thus remote weapon technologies may indirectly assist decision making as operators are not distracted by deliberation on the emotionally salient and uncertain battlefield.

UWS may also reflect the first step in increasing ethical considerations during conflict by completely removing humans from the frontline. Proponents of unmanned technologies argue that transferring lethal targeting responsibilities to autonomous systems can improve the discrimination of targets (Arkin, 2007; Schulzke, 2011). This is because they are free from cognitive load limitations and can develop almost immediate situation awareness by rapidly amalgamating a variety of disparate sensory information. For example, autonomous systems have been developed that can selectively sense and recognise enemy tanks (Dougherty, 2010). It is worth noting that the potential for *autonomous* systems is separate to the practicality of *remote* systems as a human must, from an ethical perspective, remain in the loop to make life and death decisions delivered by remote systems. Yet, arguably, the increased use of autonomous systems in the future may make strikes on non-human targets free from human error.

In summary, the following may usefully provide criteria by which one might judge the operational and ethical appropriateness of UWS during conflict: (i) the ability to define and detect a civilian; (ii) the capacity to quantify collateral damage relative to military gain; and (iii) the superiority of the system over manned human cognition. Kurkcu *et al.* (2012) analysed the impact of UWS on cognitive demands and operation tempo and advised against their use in unplanned, immediate, multi-player or situation assessment-dependent missions, due to their potential for self-jamming (i.e. when electronic attacks interfere with UWS' own electronic communications with command and control), poor sensor discrimination, and an inability to respond to requests for specified information. This suggests that human cognition needs to remain central in the design, development, and utilisation of UWS if they are to be effective; leaving aside any ethical considerations the technology is simply not advanced enough to devolve life and death decisions to. Importantly, if UWS make military decision making less just, then their potential advantages are outweighed by the potential disadvantages caused by their misapplication.

Post bellum *(after war)*

So far, we have discussed the potential ethical implications of UWS on decision making before and during conflict. It is also important to consider the ultimate *post bellum* consideration of whether actions taken in the present may enhance or hinder superordinate long-term goals. The two theoretical counter-hypotheses now discussed are whether: (i) UWS are counterintuitive to long-term strategic goals and perpetuate conflict; or alternatively (ii) UWS facilitate long-term strategic goals and facilitate peace.

The ultimate goal of any conflict is to reach a resolution and establish peace. This *post bellum* consideration underpins the superordinate aim of 'population-centric' warfare (i.e. whoever wins support of the local population will ultimately win the war) (Gentile, 2009). Population-centric approaches to war are advantageous (Kilcullen, 2006) and offer counterinsurgency strategies for gaining local civilian support (Umit and Aydinli, 2003). For example, international forces operating in Libya in 2012 were able to work effectively alongside and with rebel forces due to their collaborative approach and, likewise, it is a strategy utilised by extremist groups such as Al-Qaeda who used the popular uprising against the regime in Yemen to gain Islamist support in the south (Binnie, 2012). An important strategic question is whether UWS enhance or inhibit population-centric strategies. If they prevent the peace-focussed, long-term strategic aim of population-centric warfare, then can their use be justified?

Very little published research has so far uncovered the direct impact of UWS on population-centric outcomes. UWS are, inevitably and by definition, unmanned, and thus they reduce the opportunity to engage with civilian populations. Singer (2009: 306) has highlighted how communicating to civilian populations during conflict is 'hugely important, to the overall "war of ideas" that we are fighting against radical groups and their own propaganda and recruiting efforts'. Indeed political communication, such as diplomatic channels and open media, are more prevalent in democratically peaceful states (Choi and James, 2008) and research on civilian support for national authorities (such as the police) have found that support increases when community values are upheld and communication is non-threatening between the authorities and the public (Merry *et al.*, 2012). The more cohesive, trusting, and secure a state is, the lower the risk of future violence and extremist behaviour (Albrecht, 2006). Communication between combatants and those civilians who are 'intervened upon' is therefore important for long-term peacekeeping goals as it helps to facilitate cohesion, trust, and security. The removal of the soldier from the front line does not align with population-centric goals as it prevents the opportunity for open engagement and knowledge exchange.

A further long-term consideration for the use of UWS is with regards to their potential for inducing anti-intervention sentiment. Singer (2009: 306) has warned how poor communication during hostilities can produce 'a cloud of anger and misperceptions' in the local community, causing otherwise apathetic civilians to be dragged into conflict reluctantly. From a Western standpoint on recent conflict, this may indirectly create a breeding ground for recruitment and radicalisation of anti-intervention and extremist support. Although there is little empirical data to explain increased extremist support with confidence, some reports associate it with a rise in the number of drone attacks. Survey data in Pakistan associated an increase use of drones with support for extremist organisations (Campbell and O'Hanlon, 2009). Of the 60 reported drone attacks in Pakistan between 2006 and 2009, it was reported that 687 civilians were killed compared to only 14 Al-Qaeda targets (Ghosh and Thompson, 2009) and it has been estimated that just 2 per cent of those killed by drones are high-value targets (Stanford/NYU, 2012). Concurrently, the past decade has seen a sharp increase in anti-intervention sentiment among the 'intervened upon' population in Pakistan (Williams, 2010). A survey on those living in the FATA region of Pakistan found that 84 per cent opposed US intervention, 76 per cent opposed drone attacks, and 49 per cent believed that drones mainly killed civilians (Bergen *et al.*, 2010). It appears that the relative novelty of UWS not only biases the decision making of those who possess these systems but further impedes understanding and facilitates resistance from those civilians who are at risk of becoming collateral damage.

Not only is negative sentiment a problem for generating support for military intervention, but in extreme cases it could increase the desire for revenge (Mayer, 2009). Individuals will often take extreme risks to defend their communities when threatened (Lakatos, 2010) which can lead to spiralling victim-avenger cycles of aggression (Schulzke, 2011). When individuals feel wronged, the desire for an understanding of why it has happened is stronger than the desire to inflict equivocal harm (Gollwitzer *et al.*, 2011). Yet if there is no way to find understanding due to, for example a lack of or poor communication between the perpetrator and the victim, then revenge becomes a more attractive option. The desire for revenge is further exacerbated by 'just-world' thinking where feelings of hate are channelled into proactive violent behaviour towards the perceived deserving party (Staub, 1996; Sternberg, 2003). When a desire for revenge is coupled with a lack of accessibility to express grievances in a constructive manner because of an inability to communicate with those causing distress, then this can facilitate destructive paths of expression (Tyler, 1994). Therefore, UWS risk an environment of festering alienation, persecution, and injustice, and as such, their very use may be counterproductive to the long-term strategic aims of the most recent wars on sectarian violence.

The inevitable alternative argument is that survey data may be biased. Williams (2010) reviewed a variety of surveys and concluded that although anti-American sentiment in response to drone attacks was high in Pakistan, those who resided within the FATA tribal targeted areas were actually more tolerant and accepted drones as a necessary measure. It appeared that the risk of drone attacks was the least-worst option when living under the constant threat of terrorist organisations and so UWS may in fact facilitate long-term superordinate goals. It has been highlighted how civilian deaths in Afghanistan in 2011 increased for the fifth consecutive year, yet 77 per cent of them were attributed to Taliban and anti-government forces (UN News Centre, February 2012). Furthermore, between 2007 and 2009 anti-intervention sentiment in Pakistan increased; however opposition to Al-Qaeda also increased from 16 to 28 per cent (Campbell and O'Hanlon, 2009). This creates a confusing picture for the current status of survey data in the 'intervened upon' regions which highlights the novice understanding of the long-term implications of UWS. In-depth empirical and validated analysis is therefore needed to extrapolate findings and explicitly investigate the direct effects of UWS on the attitudes and perceptions of the 'intervened upon'. This is an integral step in evaluating the use of UWS to assess whether they are consistent with transcendent, long-term strategic goals, and therefore just tools for conflict.

Conclusions and implications

This chapter considers that the level of understanding that exists on the impact of UWS on strategic, operational, and tactical military decision making. Three pairs of theoretical hypotheses (Table 10.1) are outlined to encourage future research to explore the impact of UWS on just wars. A combination of think-tanks, technologists, expert-opinions, and psychological experiments, utilising methods such as cognitive task analysis and immersive simulated learning environments (Alison *et al.*, 2013), could help to unpack these theoretical hypotheses and guide future policy regarding their use. For example, there are restrictive policies in place for other types of weapons which limit or ban their use due to the potential for indiscriminate force (CCWC, 2001), such as non-detectable fragments (Protocol I) and mines (Protocol II). It is important to test how UWS influence conflict to ensure their appropriate use. It has already been suggested that UWS should be restricted to pre-planned, non-human targets, and that their use during time sensitive targeting is inappropriate (Kurkcu *et al.*, 2012), and there has been an increased effort over recent years to assess the implications of UWS objectively. For example, in October 2012 the UK Parliament set up an All Party Parliamentary Group on Drones to explore the domestic, international, military, and civilian use of unmanned systems (APPG, 2012); and in January 2013 the UN launched an inquiry into the human rights implications and civilian impact of the use of drones and targeted killing (UN, 2013).

As for the long-term consequences of digitalised warfare on those in the intervened upon communities, psychological research needs to be conducted with these affected communities 'in the field' in a valid, empirical, and objective way; utilising methods such as semi-structured interviewing, validated attitude scales, and coding practices with inter-rater reliability (i.e. coding by multiple researchers to remove the potential for biased conclusions). This is a challenge faced by researchers operating in conflict zones due to issues surrounding ethics and informed consent, the risk to the researcher, and also limitations in funding (Gagliardone and Stemlau, 2008). Yet it is vitally important to conduct studies to ensure that the short and medium-term tactical advantage of UWS does not come at the expense of longer-term strategic goals. An integral push in such research to explore the hypotheses outlined in this chapter will assist the international community in evaluating the effectiveness of new technologies and further in understanding when and, importantly, when not to utilise UWS in combat.

References

Akturk, A.O. and Sahin, I. (2011) 'Literature Review on Metacognition and its Measurement', *Procedia Social and Behavioral Sciences*, 15: 3731–36.

Albrecht, H.-J. (2006) 'Regaining Trust and Confidence in Post-Conflict Societies as a Way to Prevent Terrorism', in Ewald, U. and Turkovic, K. (eds), *Large-Scale Victimisation as a Potential Source of Terrorist Activities: The Importance of Regaining Security in Post-Conflict Societies*, Amsterdam: IOS Press, pp. 30–53.

Alison, L., van den Heuvel, C., Waring, S., Power, N., Long, A., O'Hara, T., and Crego, J. (2013) 'Immersive Simulated Learning Environments for Researching Critical Incidents: A Knowledge Synthesis of the Literature and Experiences of Studying High Risk Strategic Decision Making', *Journal of Cognitive Engineering and Decision Making*, 7, 3: 255–72.

Anderson, C.J. (2003) 'The Psychology of Doing Nothing: Forms of Decision Avoidance Result from Reason and Emotion', *Psychological Bulletin*, 129: 139–67.

APPG (All Party Parliamentary Group on Drones) (2012) Blog, http://appgondrones.wordpress.com (accessed 24 December 2013).

Arkin, R. (2007) *Governing Lethal Behaviour: Embedding Ethics in a Hybrid Deliberate/Reactive Robot Architecture*, Atlanta, GA: Georgia Institute of Technology.

Arkin, R. (2009) *Governing Lethal Behavior in Autonomous Systems*, Boca Raton, FL: CRC Press.

Asaro, P.M. (2006) 'What Should We Want from a Robot Ethic?', *International Review of Information Ethics*, 6: 9–16.

Bandura, A. (1986) '*Social Foundations of Thought and Actions: A Social Cognitive Theory*', Englewood Cliffs, NJ: Prentice Hall.

Baron, J. and Ritov, I. (1994) 'Reference Points and Omission Bias', *Organizational Behavior and Human Decision Processes*, 59: 475–98.

BBC News (11 October 2012) *Hezbollah Admits Launching Drone over Israel*, www.bbc.co.uk/news/world-middle-east-19914441 (accessed 15 October 2012).

Bergen, P., Doherty, P., and Ballen, K. (2010) *Public Opinion in Pakistan's Tribal Region*, New American Foundation and Terror Free Tomorrow: Washington.

Binnie, J. (2012) 'AQAP's Yemeni Takeover', *Jane's Terrorism and Insurgency Monitor*, January: 10–13.

Campbell, J. and O'Hanlon, M. (October 2009) 'Pakistan Index: Tracking Variables of Reconstruction and Security in Pakistan', www.brookings.edu/foreign-policy/pakistan-index.aspx (accessed 29 November 2011).

CCWC (Certain Conventional Weapons Convention) (2001) *Amendment to the Convention on Prohibitions or Restrictions on the Use of Certain Conventional Weapons which may be Deemed to be Excessively Injurious or to Have Indiscriminate Effects (with Protocols*

I, II and III), www.unog.ch/80256EDD006B8954/(httpAssets)/8AA7EEE69B538760C12571DE0066CC3E/$file/Amended+Article+1+authentic+text+ch_XXVI_2_cp.pdf (accessed 6 January 2014).

Chan, S. (2012) 'Loss Aversion and Strategic Opportunism: Third-Party Intervention's Role in War Instigation by the Weak', *Peace and Change*, 37, 2: 171–94.

Choi, S.W. and James, P. (2008) 'Civil-Military Structure, Political Communication, and Democratic Peace', *Journal of Peace Research*, 45, 1: 37–53.

Daddis, A. (2004) 'Understanding Fear's Effect on Unit Effectiveness', *Military Review*, July/August: 22–27.

Detert, J.R., Treviño, L.K., and Sweitzer, V.L. (2008) 'Moral Disengagement in Ethical Decision Making: A Study of Antecedents and Outcomes', *Journal of Applied Psychology*, 93, 2: 374–91.

DiBonaventura, M.C. and Chapman, G.B. (2008) 'Do Decision Biases Predict Bad Decisions? Omission Bias, Naturalness Bias and Influenza Vaccination', *Medical Decision Making*, 28, 4: 532–9.

Dougherty, M. (2010) *Modern Air-Launched Weapons*, London: Amber Books.

Evans, M. (eds) (2005) *Just War Theory: A Reappraisal*, Edinburgh: Edinburgh University Press.

Gagliardone, I. and Stremlau, N. (2008) 'Public Opinion Research in a Conflict Zone: Grassroots Diplomacy in Darfur', *International Journal of Communication*, 2: 1085–1113.

Gentile, G.P. (2009) 'A Strategy of Tactics: Population-Centric COIN and the Army', *Parameters*, 39, 3: 5–17.

Gertler, J. (2012) *U.S. Unmanned Aerial Systems*, Congressional Report N. R42136, Washington, D.C.: Library of Congress Congressional Research Service, www.fas.org/sgp/crs/natsec/R42136.pdf (accessed 13 May 2014).

Ghosh, B. and Thompson, M. (June 2009) 'The CIA's Silent War in Pakistan', *Time*, 173, 21: 38–41.

Gollwitzer, M., Meder, M., and Schmitt, M. (2011) 'What Gives Victims Satisfaction when they Seek Revenge?', *European Journal of Social Psychology*, 41, 3: 364–74.

Goodman, M. (2011) 'Killer Apps: The Revolution in Network Terrorism', *Jane's Intelligence Review*, 23, 7: 14–19.

Grossman, D. (1995) *On Killing: The Psychological Cost of Learning to Kill in War and Society*, Boston: Little Brown.

Hastie, R. (2001) 'Problems for Judgment and Decision Making', *Annual Review of Psychology*, 52: 653–83.

Holmes, R. (2003) *Acts of War: The Behaviour of Men in Battle*, London: Cassell.

Human Rights Watch (2008) *Troops in Contact: Airstrikes and Civilian Deaths in Afghanistan*, New York: Human Rights Watch.

Inman, J. and Zeelenberg, M. (2002) 'Regret in Repeat Purchase versus Switching Decisions: The Attenuating Role of Decision Justifiability', *Journal of Consumer Research*, 29: 116–28.

Joint Service Publication 383 (2004) *Joint Services Manual of the Laws of Armed Conflict*, www.mod.uk/NR/rdonlyres/82702E75-9A14-4EF5-B414-49B0D7A27816/0/JSP3832004Edition.pdf (accessed 15 October 2012).

Kahneman, D. (2003) 'Autobiography', in Frangsmyr, T. (ed.) *Les Prix Nobel 2002* [Nobel Prizes 2002], Stockholm: Almqvist and Wiksell International.

Kahneman, D. and Frederick, S. (2002) 'Representativeness Revisited: Attribute Substitution in Intuitive Judgment', in Gilovich, D. Griffin and D. Kahneman (eds) *Heuristics and Biases: The Psychology of Intuitive Judgment*, New York: Cambridge University Press, pp. 49–81.

Kahneman, D. and Klein, G. (2009) 'Conditions for Intuitive Expertise: A Failure to Disagree', *American Psychologist*, 64, 6: 515–26.

Kahneman, D. and Tversky, A. (1979) 'Prospect Theory: An Analysis of Decisions under Risk', *Econometrika*, 47: 263–91.

Kahneman, D. and Tversky, A. (1984) 'Choices, Values, and Frames', *American Psychologist*, 39: 341–50.

Kellogg, R.T. (1995) *Cognitive Psychology*, Thousand Oaks, CA: Sage.

Kilcullen, D. (2006) 'Counterinsurgency Redux', *Survival*, 48, 4: 111–30.

Klein, G. (1998) '*Sources of Power: How People Make Decisions*', Cambridge, MA: MIT Press.

Knecht, T. and Weatherford, M.S. (2006) 'Public Opinion and Foreign Policy: The Stages of Presidential Decision Making', *International Studies Quarterly*, 50: 705–27.

Kopylov, I. (2009) 'Choice Deferral and Ambiguity Aversion', *Theoretical Economics*, 4, 2: 199–225.

Kruglanski, A.W. and Thompson, E.P. (1999) 'The Illusory Second Mode, or the Cue is the Message', *Psychological Inquiry*, 10, 2: 182–93.

Kurkcu, C., Erhan, H., and Umut, S. (2012) 'Human Factors concerning Unmanned Aircraft Systems in Future Operations', *Journal of Intelligent Robot Systems*, 65: 63–72.

Lakatos, A. (2010) 'War, Martyrdom and Suicide Bombers: Essay on Suicide Terrorism', *International Journal of Axiology*, 7, 2: 171–80.

Lindlaw, S. (2008) 'UAV Operators Suffer War Stress', *Air Force Times* (8 August 2008), www.airforcetimes.com/news/2008/08/ap_remote_stress_080708 (accessed 30 July 2011).

Lord, C.G., Ross, L. and Lepper, M. (1979) 'Biased Assimilation and Attitude Polarization: The Effects of Prior Theories on Subsequently Considered Evidence', *Journal of Personality and Social Psychology*, 37, 11: 2098–109.

Mayer, J. (October 2009) 'The Predator War: What are the Risks of the CIA's Covert Drone Program?' *The New Yorker*, www.newyorker.com/reporting/2009/10/26/091026fa_fact_mayer (accessed 20 January 2012).

Merry, S., Power, N., McManus, M., and Alison, L. (2012) 'Drivers of Public Trust and Confidence in Police in the UK', *International Journal of Police Science and Management*, 14, 2: 118–135.

Muller, C. (2011) 'The Economics of Terrorism from a Policy-Maker's Perspective', *Defence and Peace Economics*, 22, 2: 125–34.

O'Keefe, M. and Wright, G. (2010) 'Non-Receptive Organisational Contexts and Scenario Planning Interventions: A Demonstration of Inertia in the Strategic Decision Making of a CEO, Despite Strong Pressure for Change', *Futures*, 42, 1: 26–41.

Olsen, O.K., Pallesen, S., and Eid, J. (2010) 'The Impact of Partial Sleep Deprivation on Moral Reasoning in Military Officers', *Sleep*, 33, 8: 1086–90.

Orend, B. (2000) 'Jus Post Bellum', *Journal of Social Philosophy*, 31, 1: 117–37.

Orend, B. (2008) 'Doctrine of Double Effect', *The Stanford Encyclopaedia of Philosophy*, http://plato.stanford.edu/archives/fall2008/entries/war (accessed 1 December 2011).

Otto, J.L. and Webber, B.J. (2013) 'Mental Health Diagnoses and Counselling among Pilots of Remotely Piloted Aircraft in the United States Air Force', *Medical Surveillance Monthly Report*, 20, 3: 3–8.

Reardon, S. (2013) 'I Spy with my Faraway Eye', *New Scientist*, 217, 26 January.

Renkl, A., Hilbert, T., and Schworm, S. (2009) 'Example Based Learning in Heuristic Domains: A Cognitive Load Theory Account', *Educational Psychology Review*, 21: 67–78.

Royakkers, L. and van Est, R. (2010) 'The Cubicle Warrior: The Marionette of Digitilized Warfare', *Ethics in Information Technology*, 12: 289–96.

Schraw, G. and Moshman, D. (1995) 'Metacognitive Theories', *Educational Psychology Review*, 7, 4: 351–71.

Schulzke, M. (2011) 'Robots as Weapons in Just War', *Philosophy of Technology*, 24: 293–306.

Sharkey, N. (2008) 'The Ethical Frontiers of Robotics', *Science*, 32, 5909: 1800–1.

Sherman, J.W., Groom, C.J., Ehrenberg, K., and Klauer, K.C. (2003) 'Bearing False Witness under Pressure: Implicit and Explicit Components of Stereotype-Driven Memory Distortions', *Social Cognition*, 21, 3: 213–46.

Singer, P. (2008) *Corporate Warriors: The Rise of the Privatised Military*, Ithaca, NY: Cornell University Press.

Singer, P. (2009) *Wired for War: The Robotics Revolution and Conflict in the 21st Century*, New York: Penguin Press.

Souchon, N., Cabagno, G., Traclet, A., Trouilloud, D., and Maio, G. (2009) 'Referees' Use of Heuristics: The Moderating Impact of Standard of Competition', *Journal of Sports Sciences*, 27, 7: 695–700.

Sparrow, R. (2007) 'Killer Robots', *Journal of Applied Philosophy*, 24, 1: 62–77.

Sparrow, R. (2009) 'Building a Better Warbot: Ethical Issues in the Design of Unmanned Systems for Military Application', *Science and Engineering Ethics*, 15, 2: 169–87.

Stanford/NYU (2012) 'Living Under Drones: Death, Injury and Trauma to Civilians from US Drone Practices in Pakistan', www.livingunderdrones.org/wp-content/uploads/2013/10/Stanford-NYU-Living-Under-Drones.pdf (accessed 6 January 2014).

Staub, E. (1996) 'Cultural-Societal Roots of Violence: The Examples of Genocidal Violence and of Contemporary Youth Violence in the United States', *American Psychologist*, 51, 2: 117–32.

Sternberg, R.J. (2003) 'A Duplex Theory of Hate: Development and Application to Terrorism, Massacres and Genocide', *Review of General Psychology*, 7: 299–328.

Strategic Defence Intelligence (2013) 'The Global UAV Payloads Market 2012–2022', *Strategic Defence Intelligence: White Papers*, www.naval-technology.com/contractors/publications-media/sdi/white-papers.html (accessed 30 January 2013).

Tversky, A. and Kahneman, D. (1974) 'Judgement under Uncertainty: Heuristics and Biases', *Science*, 185: 1124–31.

Tversky, A. and Kahneman, D. (1992) 'Advances in Prospect Theory: Cumulative Representation of Uncertainty', *Journal of Risk and Uncertainty*, 5: 297–323.

Tyler T.R. (1994) 'Psychological Models of the Justice Motive: Antecedents of Distributive and Procedural Justice', *Journal of Personality and Social Psychology*, 67: 850–63.

Umit, O. and Aydinli, E. (2003) 'Winning a Low Intensity Conflict: Drawing Lessons from the Turkish Case', *Review of International Affairs*, 2, 3: 101–21.

UN 2013 (24 January 2013) *Statement by Ben Emmerson, UN Special Rapporteur on*

Counter-Terrorism and Human Rights Concerning the Launch of an Inquiry into the Civilian Impact, and Human Rights Implications of the Use of Drones and Other Forms of Targeted Killing for the Purpose of Counter-Terrorism and Counter-Insurgency, United Nations Human Rights: Office of the High Commissioner, www.ohchr.org/Documents/Issues/Terrorism/SRCTBenEmmersonQC.24January12.pdf (accessed 30 January 2013).

van den Heuvel, C., Alison, L., and Crego, J. (2012) 'How Uncertainty and Accountability can Derail Strategic "Save Life" Decisions in Counter-Terrorism Simulations: A Descriptive Model of Choice Deferral and Omission Bias', *Journal of Behavioral Decision Making*, 25: 165–87.

van den Heuvel, C., Alison, L., and Power, N. (2012) 'Coping with Uncertainty: Police Strategies for Resilient Decision-Making and Action Implementation', *Cognition, Technology and Work*, 16, 1: 25–45.

von Neumann, J. and Morgenstern, O. (1947) *Theory of Games and Economic Behavior*, Princeton, NJ: Princeton University Press.

White, B. and Frederiksen, J. (2005) 'A Theoretical Framework and Approach for Fostering Metacognitive Development', *Educational Psychologist*, 40, 4: 211–23.

Williams, B. (2010) 'The CIA's Covert Predator Drone War in Pakistan, 2004–2010: The History of an Assassination Campaign', *Studies in Conflict and Terrorism*, 33: 871–92.

Yates, J.F. (2003) *Decision Management*, San Francisco: Jossey-Bass.

11 Keeping the human 'in-the-loop'

A qualified defence of autonomous weapons

Alex Leveringhaus and Tjerk de Greef[1]

Introduction

The use of unmanned weapons systems by western militaries has recently attracted the attention of the public and politicians. This is in no small part due to the use of unmanned aerial vehicles, also colloquially known as 'drones', by the United States in operations against alleged terrorists affiliated with Al-Qaeda (Benjamin, 2013). Drones are remote controlled weapons in the sense that, although they are unmanned, they are controlled by a human operator – or group of operators – via a satellite link. This enables operators to apply force to a target, while being located thousands of miles away. Understandably, the ever increasing distance between operators and weapons systems, as well as the ability to kill, as one commentator puts it, 'via remote control', has unnerved people (Sharkey, 2010; Strawser, 2013).

In this chapter, however, we do not want to focus on the emerging debate over remote-controlled systems as such. Instead, we are keen to point out that remote-controlled weapons are merely the tip of the iceberg when it comes to developments in contemporary military technology. More precisely, existing drone platforms offer a rough blueprint for unmanned weapons systems that will not depend on the direct presence of an operator in the actual targeting process. These weapons are autonomous in the sense that, as one definition puts it, 'they can, once activated, select and engage targets without further intervention by an operator' (Department of Defense, 2011: 13). For instance, the Taranis stealth aircraft, currently being developed by BAE Systems, can, once programmed for a particular mission, fly into enemy territory, track radar stations, and destroy these without further intervention by an operator. The Lockheed X-47B is another drone that can take off and land on an aircraft carrier, arguably the most difficult manoeuvre in military aviation, without direct guidance from an operator. While the Lockheed X-47B does not yet carry a payload, the step towards a fully weaponised system is a small one.

The debate over autonomous weapons is still in its infancy, yet it is already possible to discern two positions. On the one hand, critics, such as the Campaign to Ban Killer Robots, call for an outright ban on autonomous weapons (Human Rights Watch, 2012).[2] On the other hand, the American roboticist Ronald Arkin (2010) and the lawyer John McGinnis (2010) have both mounted sophisticated defences of autonomous weapons. In this chapter, we (a philosopher and a cognitive engineer) want to engage critically with the debate about autonomous weapons. We begin by tackling some conceptual issues. We discuss the most important arguments in favour of autonomous weapons and examine two objections to them. Finally, we develop a perspective on autonomous weapons that calls for the introduction of autonomous elements into existing systems, rather than the creation of fully autonomous systems.

Autonomy and targeting systems

To get a good grasp of the main issues in the debate over autonomous weapons, it is necessary to disambiguate central concepts. In this part of the chapter, we will first outline what we understand by the term 'weapon' and weapons system; we will then engage with the concept of autonomy. Before we begin, we would like to raise two general points. First, although just war theory, whose insights will play an important role later in the paper, has witnessed a remarkable renaissance in philosophical circles, philosophers, with few exceptions, have paid little attention to the conceptual issues involved in many weapons systems (see Finnis *et al.*, 1988; Strawser, 2013). To be sure, philosophers have assessed how certain weapons relate to the central just war criteria of distinction, proportionality, and necessity. But more conceptual work is required to grasp the distinctiveness of specific types of weapons and the different technological elements embedded in them. We hope that the chapter can serve as a step towards a broader philosophical debate about the technological transformation of armed conflict in the twenty-first century.

Second, the discussion in this chapter is limited to what we refer to as *targeting systems*. There are two reasons for our focus on these types of systems. The first is that while targeting systems will share certain features with other types of military technology, they pose, as we will show, some distinctive ethical challenges that require further analysis. The second reason is more pragmatic. The kinds of autonomous weapons that will be developed over the next couple of decades will largely be targeting systems, or at least be integrated into existing weapons systems. Because of this, we want to subject targeting systems to greater critical scrutiny. Without a clear understanding of what targeting systems are, it will be difficult to see why the issue of autonomy is proving so controversial.

Systems

Like other weapons systems, targeting systems are *systems* in the sense that there are set procedures for decision making, involving a number of pre-determined parameters. Targeting decisions, in other words, are not made on an arbitrary basis. Rather, decisions reflect the regulatory framework that restricts the use of military force. No weapons *system* exists in a legal or moral vacuum. It is usually embedded within a broader framework of rules and regulations. This point, we argue, is not always fully appreciated in the debate over autonomous weapons. We will return to it in our fourth section.

It is noteworthy, though, that the mere fact that decisions are not made arbitrarily does not guarantee that these decisions are correct. Faulty intelligence, for instance, can lead to disastrous outcomes, even though the decision making criteria underlying the system have been applied correctly. Also, given that human beings are fallible creatures, targeting criteria may be misapplied. Obviously, systematic targeting, though it may sometimes lead to mistakes, is morally preferable to arbitrary targeting. After all, as we shall see in section 3, the use of force during war is regulated by a number of important legal and normative criteria. At the same time, it is crucial to acknowledge that no system is perfect. Bearing this in mind, it is important to minimise 'system failures'. We will return to this point in section 5.

Targeting systems

While targeting systems share many features with other military systems, they are also distinctive in the sense that they are designed specifically to apply kinetic force to a target in order to destroy it. Conceptually, then, targeting systems differ from weapons systems, such as electronic countermeasure systems, which are not designed specifically to destroy objects directly. Admittedly, the use of an electronic countermeasures system can have foreseen yet unintended side-effects that lead to the destruction of an object. Imagine Blue, in the course of fighting a defensive war against Red, blocks the communications system of Red's air force. Because of this, some of Red's airborne fighter jets might crash. One could argue that Blue merely foresees but does not intend the crashes. In order to ensure that Red's air force cannot continue its offensive, Blue only needs to block Red's communications system.[3] Blue does not need to destroy any of Red's planes. According to our definition, then, Blue's electronic countermeasures system is classifiable as a weapons system but not a targeting system. To be sure, the use of targeting systems can also have unintended side-effects. But targeting systems are morally and legally problematic precisely because they exist *in order* to kill or destroy.

Autonomy

In the present context, relevant targeting systems include, but are not limited to, military robots, such as drones, as well as non-robotic computer-based systems, such as the Israeli Iron Dome missile defence system. The distinction between robotic and non-robotic (targeting) systems is not clear cut, but we do not want to enter into a debate about this issue here. Instead, we are keen to stress that both types of targeting system are classifiable as cognitive systems. This means that, depending on the quality of their software algorithms, they are potentially capable of (1) acquiring information about an environment via sensors, (2) analysing the information, (3) making a decision about taking a specific course of action, and (4) executing the decision. Each of the four phases of this information processing loop can be automated at different levels, which means that the tasks associated with a particular phase are carried out, to a greater or lesser degree, by the system itself. Traditionally, targeting systems have relatively high levels of automation in phases (1) and (4), but not phases (2) and (3). Generally, their status as cognitive systems indicates why computer-based and robotic targeting systems are considered to be potentially autonomous.

That said, let us note that the use of the term autonomy appears odd, especially in a military context. Playing a major role in contemporary moral and political philosophy, autonomy denotes that an agent is capable of pursuing a conception of the good life based on his/her own reasons and motives, rather than those of another agent (Kant, 1998; Rawls, 1999a). For those who believe that autonomy is an important moral good, the institution of the military is morally problematic because, in order to function, it needs to diminish, though not extinguish, the autonomy of its members. As the eminent just war theorist Michael Walzer succinctly puts it, 'war is not a relation between persons but between political entities and their human instruments' (Walzer, 2006: 36). Now, while it can be permissible to use human persons as instruments in certain circumstances, Walzer's comment indicates that, in times of war, members of the military typically do not act out of their own motives but the motives of the political entity – the state – of which they are members. The tension between individual autonomy and the structures of authority and obedience that ensure the functioning of the military as an instrument at the disposal of the state has characterised many recent debates within just war theory (McMahan, 2009). It is beyond the scope of this chapter to assess these debates. It suffices to note that, when considered against this background, it is strange that the military should be interested in developing weapons that are specifically designed to be autonomous.

The solution to this riddle is that the term autonomy is used differently by the philosophical and the engineering communities. To avoid confusion, we deem it necessary to differentiate between moral autonomy and what we call operational autonomy. The latter type of autonomy, with which we are concerned in this chapter and in which many researchers in engineering are also concerned, denotes that a machine is capable of carrying out specific tasks without a human operator and without being subject to certain regulatory constraints. More precisely, in cognitive engineering, the concept of operational autonomy is used in two ways.

1 Self-sufficiency (or technological capacity): The term operational autonomy is used to signal that a machine is self-sufficient. Self-sufficiency, as it is understood here, relates to the ability of a machine to carry out certain tasks without assistance from an operator. Cognitive engineers define different levels of self-sufficiency for each phase of the aforementioned information processing loop. A machine may enjoy high self-sufficiency with regard to all four levels of the loop or only some of them;
2 Self-direction: The term operational autonomy is also used to signal that a machine is self-directing. The concept of self-direction, in the present context, refers to the set of actions a machine is allowed to perform. Low self-directedness indicates that a machine may be technologically capable of performing certain actions, but for legal, moral, or political reasons is not allowed to do so. By contrast, high self-directedness indicates that there are only minimal restrictions in place to regulate the conduct of the machine. As a result, the machine has far reaching authority over its actions. Again, self-direction needs to be viewed against the background of the four stages of the information processing loop. To use a contemporary example, the Rules of Engagement state that jet fighter pilots are not allowed to use the jet's fire radar system during the first phase of the information processing loop. During subsequent phases, though, this restriction might be lifted. Similarly, operationally autonomous machines will be subject to specific restrictions at different stages of the information processing loop.

Without going into detail, let us note two points about operational autonomy. First, the difference between moral and operational autonomy is normatively significant. It is relatively uncontroversial that morally autonomous agents can be held responsible for their actions. After all, 'their' actions are truly *theirs*; they are not determined by someone else. The fact that a machine is operationally autonomous, on the other hand, does not imply that it is morally or legally responsible for its actions. While morally autonomous individuals are moral agents who can be praised and blamed for *their* actions, operationally autonomous machines are not moral agents. Even high levels of operational autonomy do not change this. Moral and operational autonomy do not exist on a continuum. There is, in other words, no point where operational autonomy becomes moral autonomy. Of course, operationally autonomous machines of the type we are currently considering also 'do' things. Operationally autonomous machines, especially if they have been equipped with learning mechanisms, may even review their performance of a particular task and adapt their behaviour in order to perform the task more efficiently or effectively. But none of this changes the fact that the task they perform is assigned to them by another agent, their operator. Expressing this thought in more philosophical language, we can say that the behaviour of operationally autonomous machines is classifiable as morally heteronomous. What the machine may or may not do is determined by someone else. Needless to say, this raises the question whether the operator, given that s/he has assigned a particular task to an operationally autonomous machine, is responsible for what the machine does. Common-sense implies s/he is. Yet, as we shall see below, this assumption has been questioned by critics of autonomous weapons systems.

Second, the concept of operational autonomy has important implications for the relationship between human individuals and machines. As the concept of self-sufficiency implies, unlike remote controlled devices, such as drones, machines with operational autonomy do not directly rely on a human operator. Consequently, while in the case of non-autonomous systems a human operator is always 'in the [decision making] loop', the operator could, in principle, be taken 'out of the loop' of an operationally autonomous system. In out-of-the-loop systems the ability of the operator to stop the system may be greatly reduced. In some cases, it may even be non-existent. Alternatively, the operator may remain 'on the loop', which means that s/he is capable of overriding the actions of an operationally autonomous machine. The distinction between 'in the loop' (remote-controlled), 'on the loop' (operator on stand-by) and 'out of the loop' systems is commonly used in the current discourse over operationally autonomous weapons, and we shall follow it here.

Taken together, the above observations should provide a clearer idea of the kinds of operationally autonomous weapons systems that we want to discuss in this chapter. Returning to the debate over autonomous weapons, we think it is fair to say that participants have in mind weapons that are (1) targeting systems, (2) wholly self-sufficient with regard to all four stages of the information processing loop (Self-sufficiency), and (3) permitted to apply force to certain targets (Self-direction). We call these types of targeting systems *fully* operationally autonomous and refer to them as FOATS (Fully Operationally Autonomous Targeting Systems). In the case of FOATS, the human operator can either be taken out of the loop completely (out-of-the-loop FOATS) or remain on stand-by in order to override the machine's decisions. Yet, since, as we indicated, operational autonomy is a matter of degree and therefore highly dependent on the quality of the governing software of a system, there can be operationally autonomous targeting systems that have less than *full* operational autonomy. We set these types of systems aside for now. With this conceptual background in place, let us now scrutinise the main argument in favour of FOATS, as well as the two main objections levelled against it.

The case for autonomous weapons and the Non-Compliance Objection

In addition to the claim that – like remote-controlled weapons – operationally autonomous weapons protect the lives of service personnel, the main morally relevant argument in favour of operationally autonomous weapons is surprisingly straightforward. Advocates of FOATS can point to the appalling track record of human decision making during armed conflict (Arkin, 2010). Especially when under pressure, humans often make disastrous decisions with regard to the delivery of force. Consider the notorious USS *Vincennes* case (1988), where an American naval vessel mistakenly shot down an Iranian civilian airliner. Due to their bias towards a particular scenario, the human operators of the USS *Vincennes*' missile targeting system consistently misinterpreted the information relayed by their instruments, thereby reinforcing their mistaken belief that a civilian aircraft operated by Iran Air was a fighter plane about to attack them (Cannon-Bowers and Salas, 1998). As a result, 290 people lost their lives. Because, as Ronald Arkin contends, machines are not susceptible to this kind of situational bias and are thus immune to psychological stress resulting from it, FOATS are likely to have interpreted the data correctly and abstained from firing at the civilian aircraft. The example illustrates what we take to be the most important argument in favour of FOATS: computers make better targeting decisions during difficult and stressful combat situations than humans. If this is true, it makes sense to take the human operator out of the loop altogether or at least reduce his role to a stand-by function. One important criticism, which we refer to as the Non-Compliance Objection, denies this. FOATS, it contends, will not make sound targeting decisions because they lack the capacity to comply with central legal and ethical principles.

To assess whether those critics who voice the Non-Compliance Objection have a point, we focus on those criteria that are directly relevant to the targeting process. These are (1) the principle of proportionality, (2) the principle of military necessity, and (3) the principle of distinction. Since our interests, in this and the next section of the chapter, are philosophical in orientation, we shall discuss the above principles as they feature in just war theory, rather than international law – though there is clearly an overlap between the two discourses. From the perspective of just war theory, the principles of distinction, proportionality, and necessity are central to the framework of *jus in bello*, which governs the conduct of hostilities. The case for or against FOATS, then, will largely be decided by their ability to comply with *jus in bello* norms.

Let us begin by analysing the relationship between the Non-Compliance Objection and the principle of distinction. The latter principle requires warring parties to distinguish between combatants and non-combatants. Without it, it is hard to see how military activity could be prevented from descending into slaughter. According to the Non-Compliance Objection, since human behaviour during armed conflict is too complex and unstructured to be intelligible to a machine, FOATS cannot distinguish between combatants and non-combatants. As a result, the delivery of force via FOATS is always potentially indiscriminate. For this reason, critics contend, FOATS should be banned. We think this observation is correct. Unless there are means to make combatants 'visible' to a machine, such as the integration of a microchip into their uniforms, it is hard to envisage how machines could discriminate between non-combatants and combatants.

That said, it is worthwhile pointing out that the scope of this aspect of the Non-Compliance Objection is limited. This is because, in some contexts, FOATS are capable of discriminating between military and non-military targets. To illustrate the point, some naval defence systems, such the Dutch Thales Goalkeeper system, are under strict conditions and in particular contexts only already capable of identifying and destroying hostile missiles with considerable accuracy (Arciszewski *et al.*, 2009). Especially when decisions concerning the delivery of force are made under pressure due to short reaction times, having a human operator in the loop may unnecessarily delay the delivery of defensive force. Therefore, it seems reasonable, in certain situations, to rely on FOATS, provided they are capable of identifying their targets reliably and accurately. If this is correct, the Non-Compliance Objection only suggests that FOATS must not be used directly against human targets.

Critics could counter-argue that the use of FOATS, even if their operations are restricted to the identification and destruction of non-human targets, will lead to the loss of human life. In this regard, it is useful to distinguish between two scenarios. First, imagine a naval defence system which can accurately identify the attack patterns of jet fighters. If, as a result, a jet fighter plane is shot down, the pilot might die. Yet we can reasonably assume that individuals operating fighter planes are classifiable as combatants. Air force pilots, after all, are highly trained military professionals. The use of FOATS, in this case, does not violate the principle of distinction.

Second, and more controversially, imagine that because the Iron Dome system destroys an incoming missile, non-combatants are harmed by falling debris. Surely, this case is more problematic than that of the fighter pilot. But just war theorists can reply as follows: according to the Doctrine of Double Effect, an often cited moral principle in just war theory, it is permissible to cause the death of non-combatants as a foreseen yet unintended side-effect of a particular act, subject to necessity and proportionality (see Cavanagh, 2006). Note that, from the perspective of orthodox just war theory, the Non-Compliance Objection only applies to cases where force is used intentionally against non-combatants. It does not, as we understand it here, apply to unintended side-effects. If this is sound, then, in the Iron Dome case, it is permissible to harm non-combatants because their death is an unintended side-effect of the destruction of the enemy missile, provided that these deaths are proportionate – an issue we shall return to shortly. The harm inflicted on non-combatants is not a means to an end. That is, it is not required to destroy the hostile missile. Assuming the Doctrine of Double Effect is defensible, the use of FOATS may be permissible, notwithstanding its impact on non-combatants.

Interestingly, the invocation of the Doctrine of Double Effect points towards a broader issue. The doctrine, as we just explained, distinguishes morally and conceptually between what an agent intends and what he foresees. In the above case, the question is whether it is morally relevant that FOATS 'foresee' the death of any non-combatants or whether the distinction between intended and foreseen actions is exclusively relevant to the actions of human agents.[4] We contend that the latter is the case. Machines, in our view, are not moral agents because they lack moral autonomy. Consequently, what matters are the intentions of the operators who deploy the machine. Operators will foresee that the activities of FOATS might cause unintended harms, in addition to intended harms. The relevance of this will become fully apparent in the next section of the chapter.

Critics could grant this point while raising doubts over the ability of FOATS to comply with the remaining criteria of *jus in bello*, necessity and proportionality. The latter criterion appears twice in *jus in bello*. On the one hand, the proportionality criterion should be viewed as an independent criterion that requires warring parties to avoid causing excessive harms. In this sense, proportionality calculations must be made with regard to any kind of harm, regardless of whether it is intended or merely foreseen. On the other hand, we find a more restricted version of the proportionality criterion contained in the Doctrine of Double Effect. In this context, the proportionality criterion demands that any unintended side-effects must not be disproportionate to the intended good effect. Consequently, according to the Doctrine of Double Effect, the performance of an act is impermissible if its unintended side-effects cause unnecessary and disproportionate harm. For our purposes, it is not necessary to enter into this and related intricacies. Rather, we want to point out that the two criteria pose, in our view, insurmountable problems for FOATS.

It might be tempting to assume that FOATS could be programmed in order to comply with both criteria. Perhaps, defenders of these systems could argue, we can invent algorithms that prevent FOATS from causing disproportionate and unnecessary harm. However, this poses two immediate problems. First, any calculations of likely harms would involve an assessment of the impact of military actions on combatants *as well as* non-combatants. The principle of distinction is thus directly relevant to the criteria of proportionality and necessity. As a rule of thumb, we argue that the interests of non-combatants should be given more weight than those of combatants (Walzer, 2006: 154–156). But, as we just saw, FOATS will not be able to distinguish between combatants and non-combatants in the first place. Therefore, they will lack a sound basis for making appropriate judgements of proportionality and necessity.

The second problem has to do with the inherent vagueness of the necessity and proportionality criteria. Even if compliance with the principle of distinction could be ensured, FOATS will find it hard to interpret what constitutes proportionate and necessary harm. More precisely, the problem is that the application of the necessity and proportionality criteria is highly context-dependent. It does not just involve a rigid application of a particular formula. This partly explains why these criteria are vague and why, to our knowledge, no soldiers have been legally prosecuted for violations of them. To illustrate the point, consider the following scenario. Blue is engaged in a self-defensive war against aggressor Red. In order to repel Red's attack, Blue needs to destroy a weapons factory in Red's territory. Suppose that, at t_1, Blue is about to lose the war and ponders whether to attack the factory. If, at t_1, Blue bombed the factory, 100 non-combatants would die as a side effect of the strike. However, assume that, because of bad weather at t_1, Blue has to call off the attack at the last minute. At t_3, weather conditions have improved and Blue considers attacking the factory again. Fortunately, however, at t_2, Blue secured a major victory on a different front. Because of this, while causing 100 non-combatant deaths might have been permissible at t_1, it may well be excessive and therefore disproportionate to cause that many deaths at t_3.

The scenario shows that one cannot assign fixed military values to potential targets. It is impossible to programme a rule into FOATS that states that targeting X-category objects causing z-amount of casualties amongst non-combatants would be excessive, while causing y-amount of casualties in targeting X-category objects would not be. This reveals, we believe, a broader point about just war theory and the various criteria it contains. The application of these criteria is not just a matter of rule following. Rather, it involves making ethical and political judgements in constantly shifting circumstances. Arguably, there is a parallel here with Aristotelian moral thought. For Aristotle, our particular moral judgements depend on acquiring wisdom and experience. Rather than following a simple rule, we should ask ourselves how a person who possesses both would act (Aristotle, 2000). Perhaps the application of individual just war criteria functions in a similar fashion. We would need to ask ourselves how an experienced military commander or an experienced politician – a true statesman – would act under the circumstances (see Rawls, 1999b). By contrast, although highly operationally autonomous machines, such as FOATS, will be able to follow rules along the lines of 'If X, then do y', they lack the capacities for a *moral judgement* that would be required to adequately apply proportionality and necessity criteria.

Of course, it is possible to develop all kinds of fanciful science fiction scenarios, perhaps taking a cue from the great works of the late Iain M.

Banks, in which humans develop machines with the capacities to make these sorts of judgements (Banks, 1992). We could imagine a world in which it becomes possible to programme the 'wisdom' and 'virtues' of a 'perfect' military commander or indeed a statesman into a machine, so that 100 per cent compliance with the rules of *jus in bello* is achieved every time a targeting decision is made.[5] We do not think that these scenarios are particularly helpful in the current debate about operationally autonomous weapons. The types of FOATS which are being developed at the moment and which are likely to be deployed over the next five to thirty years will not resemble the highly sophisticated, almost God-like, machines we can imagine in philosophical thought experiments. But just as the development of remote-controlled systems has had an impact on what we understand by modern warfare, increasing automation within the military will challenge our thinking about armed conflict in the twenty-first century. We need to respond to these challenges now, rather than invent potential plotlines for the next Terminator movie.

Overall, then, if the above observations are correct and we focus on those systems that are realistically going to be developed within the next two to three decades, the Non-Compliance Objection has considerable force. That said, the scope of the objection, we think, is restricted. It is one thing to say that FOATS will find it difficult to apply the usual *jus in bello* criteria themselves. It is quite another to say that it will be impossible for human operators to comply with *jus in bello* rules when considering whether or not to deploy an operationally autonomous system. The context in which FOATS are about to be deployed is key here. At the current stage of technological development, it is hard to see how FOATS could be used safely in an ordinary battlefield context where they might encounter or interact with humans. But the use of FOATS during blue water naval operations systems may potentially meet the standards set by *jus in bello*. Blue water environments are very different from urban combat environments. Thus, as our discussion of the principle of distinction showed, FOATS can be used in cases where the signatures of potential targets are relatively unambiguous and stable. Although the Non-Compliance Objection has considerable force, it is not sufficient to justify a blanket ban of all types of FOATS. Still, the above points indicate that the development of genuine out-of-the-loop systems, where the operator is not on stand-by, is a bad idea. Operators should be able to block applications of force that may lead to disproportionate and/or unnecessary harm.

Before we discuss a further objection to FOATS, we would like to raise a final point. We believe that fairly generalised judgements about necessity and proportionality can and should be factored into the design of FOATS. While it is impossible to plan for all eventualities, designers should take into account how FOATS are likely to impact on non-combatants in the contexts in which they might be deployed. Based on their analysis, designers should work hard to restrict the harmful side-effects of FOATS as much as possible. For instance, in order to predict whether the use of FOATS has disproportionate side-effects, engineers and military officials should consider integrating a debris fallout prediction model into the system. Because some side-effects are foreseeable, they can be subjected to further analysis by an appropriate computer system, which, in case of any anomalies, could alert an operator on stand-by.

Part of our defence of FOATS rests on the assumptions that these systems, if they are to be developed at all, should be on-the-loop systems in which an operator can block applications of force or otherwise override the system. However, some critics of FOATS contend that the role of the operator becomes morally problematic once an operationally autonomous system has been deployed. This takes us to the second objection to FOATS, which we call the Abdication of Responsibility Objection.

The Abdication of Responsibility Objection

Unlike the Non-Compliance Objection, the Abdication of Responsibility Objection does not focus on the quality of targeting decisions and the capacity of FOATS to comply with the principles of *jus in bello*. Rather, it contends that no one can be held responsible for targeting decisions made by FOATS. The philosopher Robert Sparrow, who states this objection with great force, fears that once humans are taken out of the loop no one can be held responsible for the actions of FOATS (Sparrow, 2007). To illustrate this point, Sparrow imagines what he calls a 'robo-war crime' in which an autonomous system kills a group of surrendering combatants because it has determined that keeping them alive would be too costly. As Sparrow rightly notes, killing those who are about to surrender is, legally and ethically, a war-crime.

The Abdication of Responsibility Objection, which emerges out of Sparrow's reflections, has two strands. The first assumes that it is inappropriate to hold a human agent, such as a software programmer or commanding officer, responsible for the 'robo-war crime'. According to Sparrow, this is because (operational) autonomy introduces an element of unpredictability into FOATS's behaviour. If FOATS engages in (some) actions that cannot be predicted, it is inappropriate, Sparrow maintains, to hold another agent responsible for those actions. The second strand of the Abdication of Responsibility Objection contends that FOATS cannot be held responsible for their actions either. This means that, in the case of the robo-war crime, the system that killed the surrendering combatants cannot be held responsible for its deeds.

While, for reasons we do not want to discuss here, Sparrow's account of the 'robo-war crime' resembles one of the fanciful science fiction scenarios we criticised above, Sparrow expresses a general worry that is sometimes voiced with regard to operationally autonomous technology. The philosopher Andreas Mathias, for instance, thinks that operational autonomy opens up what he calls a 'responsibility gap', in which no-one can be held responsible for what a machine does, precisely because what the machine does is not entirely predictable (Mathias, 2004). For Mathias, the unpredictability of operationally autonomous machines results from the use of learning mechanisms within the system. To explain, in order to navigate a complex battle space or carry out sophisticated tasks, FOATS will be equipped with some learning mechanisms that enable the machine to adapt to its environment. This is necessary because it is impossible for programmers to foresee every eventuality and program FOATS accordingly. Doing so would take decades. The interaction between an operationally autonomous machine, its learning mechanisms, and its environment will lead to outcomes that are not fully predicable. This is, in a nutshell, the reason why Mathias thinks we are faced with a responsibility gap.

Rather worryingly, if one imagines too many of these gaps opening up in an organisation we will arrive at state of affairs where a commitment to responsibility and accountability is undermined. This is why we refer to the responsibility-centred criticisms of FOATS as the Abdication of Responsibility Objection. Surely, the loss of a commitment to responsibility and accountability in our social and political institutions, especially the army, cannot be in anyone's interest. But is the Abdication of Responsibility Objection sound? We do not think it is. Before we outline the reasons for our scepticism, we want to briefly indicate where we agree with Sparrow and Mathias.

Sparrow, in particular, is correct in arguing that it does not make sense to hold a computer responsible for its actions. This is seriously considered by the philosopher Daniel Dennett – but note that Dennett's, as well as Sparrow's observations, pertain to systems that are immeasurably more sophisticated than the ones we are considering at the moment (Dennett, 1997). Nevertheless, we concur with Sparrow that it does not make sense to hold a computer responsible for what it does, albeit for slightly different reasons. The point, for us, is that we do not *need* to hold a computer responsible. One important justification, usually given by consequentialist moral theories, for holding human individuals responsible is that, by praising or blaming them, we want to somehow affect their behaviour (Brandt, 1992: 215–34). Naturally, there are limits to what we may do in order to influence an individual's behaviour. Cruel and degrading forms of treatment are certainly ruled out, even though they might ensure high-levels of compliance in the future. But the same restrictions do not apply in the context of operationally autonomous machines. If we want to suppress undesirable behaviour in an autonomous machine, we can simply send it an error message, switch it off, or even re-programme it. It goes without saying that it would be unethical to re-programme humans, even if we were able to do so.

Of course, it is one thing to say that a machine is not responsible for what it does; it is quite another to say that humans who rely on operationally autonomous machines cannot be held responsible for violations of legal and moral codes that result from the deployment of a particular machine. Does the inherent unpredictability of the behaviour of operationally autonomous machines present a sound reason that could be cited by human operators in order to be exculpated from wrongdoing?

Our first answer to this question is a direct response to Sparrow's 'robo-war crime' scenario. The decision to kill the surrendering combatants in order to avoid excessive costs cannot, contrary to Sparrow's argument, arise solely within the machine. Someone, either at the design phase or the mission programming phase (see below), must have given the machine permission to relate lives to costs, and make targeting calculations based upon these two criteria. This follows from the concept of Self-direction, which, as we argued above, forms an important part of our conception of operational autonomy. In allowing the machine to make these types of calculations, the operator must be aware of the risks this may pose. The operator is obliged to prevent the machine from relating the two targeting criteria in ways that could result in a war crime. The fact that the operator did not foresee that the machine would engage in a 'robo-war crime' in a *particular* instance does not mean the operator cannot be held responsible for said crime. The operator could have foreseen that there may be situations in which the machine makes inappropriate calculations. The question is whether the operator, in deploying the machine and allowing it to relate lives to costs, violated a certain standard of care. If they did, they can be held responsible for the 'robo-war crime'.

Second, and directly related to the preceding point, unpredictability is a matter of degree. As a rule of thumb, we can say that, in a military context, unpredictability is relatively low in the information acquisition and action phases of the targeting process. It is higher and more problematic in the following analysis and orientation phases. There is a danger that FOATS will not be able to analyse potential targets accurately and, therefore, fail to orient their actions properly. Our response to this problem is that the use of FOATS should be restricted to environments where the likelihood of unpredictable behaviour is low and the possibility of flawed analytical processes is minimal. Surely, operators can be held responsible for deploying systems in contexts where doing so is unreasonably risky or where the results of unpredictable behaviour would have catastrophic consequences.

Third, one should not forget that FOATS are usually part of an organisation. As the concept of Self-Direction implies, the use of FOATS is always subject to the rules and regulations governing a specific organisation, in our case the military. The rules of a military organisation, for instance, determine which functions FOATS are permitted to utilise during a mission. In fact, it is likely that FOATS will be specifically programmed for each mission. Note that this point reflects what we said about the nature of the military in section 2. The relationships of hierarchy and obedience obtaining within the military make it necessary that the latter aims to suppress 'autonomous' behaviour in its members. It seems odd to assume that the military would want to develop weapons that it cannot control, given how much emphasis is usually placed on the principle of 'command and control' in military institutions. In general, the fact that FOATS are always embedded within an organisation also indicates who is responsible in case things go wrong. Ultimately, the organisation that chooses to deploy FOATS is responsible. In cases of 'wrongdoing' by FOATS, for example, the organisation might be liable to pay compensation to the victims. Depending on the case, the organisation might also be subject to criminal prosecution via the international legal system. Arguably, international (criminal) law might have to be adapted to deal with such cases. But this will be the topic of another paper.

If these points are sound, the Abdication of Responsibility Objection does not succeed in showing that the development and deployment of all types of FOATS should be made illegal. Overall, then, we think that the Non-Compliance and Abdication of Responsibility Objections are neither individually nor jointly sufficient to justify a blanket ban of FOATS. Nevertheless, some concerns remain with regard to operationally autonomous weapons technology. At the moment, the contexts in which FOATS can be safely deployed seem fairly limited. In the final section of the chapter, we want to suggest that, rather than concentrating on *full* operational autonomy, we should instead strengthen operationally autonomous elements *within* combat systems.

Operational autonomy *in* combat systems: a design proposal

If anything, critics and advocates of FOATS identify the strengths and weaknesses of humans as well as computer-based cognitive systems, respectively. Humans frequently get it wrong, but machines are not always better. It is a common sense response, we think, to try and 'team up' human operators and computers (artificial agents) by integrating them into a *joint* cognitive system (Hollnagel and Woods, 2005). The strengths of one actor can then compensate for the weaknesses of the other. This move, we contend, potentially enhances situational understanding and subsequent decision making. The notion of cooperation between human agents and artificial agents requires a paradigm shift from FOATS to e-Partnership-based systems. To avoid potential misunderstandings, the concept of an 'e-Partnership' should only be understood metaphorically. The idea, especially in cognitive engineering, is that it is useful to view machines as analogous to a junior team member in order to better integrate it into existing work processes. But this does not change the fact that the machine is not a moral agent and, as a result, cannot be an equal party to a 'partnership'.

The idea of an e-Partnership is merely a useful metaphor in order to optimise working processes – nothing more, nothing less.

Admittedly, the concept of e-Partnerships is not entirely new. E-Partnerships are already being prototyped in the domains of health care, space missions, and naval warfare (see de Greef *et al.*, 2010) For now, however, we are content with briefly outlining how e-Partnerships might work in the context of targeting systems. Generally, e-Partnerships are facilitated by what cognitive engineers refer to as 'working agreements' (de Greef *et al.*, 2010; Miller and Parasuraman, 2007). Again, the concept of a working agreement should only be viewed metaphorically; it should not be taken literally. In the working agreement, the operator sets out the particular tasks that the computer should work on. In the context of targeting systems, working agreements give rise to a fine-grained division of labour with regard to specific tasks associated with the four stages of the information processing loop, and in relation to particular objects of interest, such as airplanes, missiles, or vessels. As part of a working agreement, for instance, a human operator may delegate the task of identifying relatively unambiguous objects to the artificial agent, while remaining in control of the more demanding task of identifying ambiguous objects.

To illustrate the point, imagine that, during the deployment of a future version of Israel's Iron Dome missile defence system, targeting decisions need to be made with regard to two cases. The first involves missiles that fit an easily identifiable attack and speed profile. In the second case, by contrast, objects are not readily identifiable as missiles but still exhibit an attack pattern. Based on a working agreement, the artificial agent might deal with the first case, thereby giving the operator the necessary room to analyse and deal with the second. This scenario is not far-fetched. In real life, a similar working agreement was tested and evaluated in a naval combat workstation prototype. The Dutch Navy officers who participated in the study highly appreciated the division of labour between human and artificial agents, especially when decisions had to be made under pressure (Arcizewski *et al.*, 2009).

Advocates of FOATS could reply that the concept of e-Partnerships does not solve the initial problem: humans are bad at decision making. But if the arguments of sections 3 and 4 of this chapter are correct, placing the operator outside of the (decision making) loop is neither always possible nor necessarily desirable. Moreover, we think that the e-Partnership approach can meet the advocates of full operational autonomy half-way. The point of e-Partnerships is to reduce the burden on the operator by allowing him or her to concentrate on important tasks. Hopefully, this reduces the stress experienced by the operator when carrying out tasks in high-risk domains. Further, depending on the nature of the e-Partnership, an(other) artificial agent could monitor some of the operator's actions and *vice versa*, thereby reducing the potential for misapplications of force.

Admittedly, though, the e-Partnership approach does not guarantee perfect decision making. But neither does full operational autonomy.

Critics of FOATS, such as the Campaign to Stop Killer Robots, should welcome the e-Partnership approach because it does not dispense with human operators. The human remains in the loop. First, by allowing the human operator to be involved in targeting decisions where the situation is unclear, it addresses the central concern of the Non-Compliance Objection, namely, that FOATS will find it difficult, if not impossible, to discriminate between combatants and non-combatants. Second, by leaving the operator in the loop and stressing the importance of working agreements, the e-Partnership approach dispels some of the worries underlying the Abdication of Responsibility Objection. In the end, human operators will be responsible for the terms of the working agreement. Third, by allowing artificial agents to monitor the actions of human operators and vice versa, e-Partnerships thus have the capacity to prevent applications of force to non-combatants. The machine, for instance, might correct flawed actions by the operator and block the application of force.

In sum, we believe that the e-Partnership approach has the potential to reorient the debate over operationally autonomous weapons. Because the entrenched positions of critics and advocates as well as the narrow focus on FOATS leave little room for future development, the debate is bound to reach a stalemate sooner or later. The idea of an e-Partnership can help us avoid such a stalemate.

Conclusion

Our analysis shows that both critics and advocates of FOATS raise important points. However, neither is capable of making an unequivocal case for or against FOATS. In this chapter, we argued that FOATS are suitable in some contexts but not others. At the moment, their deployment is clearly limited to defence against enemy missiles or other types of air power-based attacks. It is inconceivable, however, that FOATS could be used in counterinsurgency operations. In this light, our defence of FOATS is necessarily qualified. Moreover, the e-Partnership approach meets some of the concerns raised by critics and advocates of FOATS. While it allows for certain degrees of operational autonomy, it leaves the human in the decision making loop. Combining the respective strengths of machines and humans, this approach has significant potential for enhancing targeting decisions.

Notes

1 We would like to thank Janina Dill, Marco Meyer, Cheney Ryan, Henry Shue, and Hugo Slim for their comments on an earlier draft of this chapter. We would also like to thank Nicola Horsburgh for her editorial assistance with later drafts of the chapter.

2 www.stopkillerrobots.org (accessed 8 January 2014).
3 This is consistent with the so-called Doctrine of Double Effect. We return to the Doctrine when we examine the first main objection to autonomous weapons.
4 We would like to thank Henry Shue for alerting us to this point.
5 We thank B.J. Strawser for his views on this issue.

References

Arciszewski, H.F.R., de Greef, T.E., and van Delft, J.H. (2009) 'Adaptive Automation in a Naval Combat Management System', *IEEE Transactions on Systems, Man, and Cybernetics Part A: Systems and Humans*, 39, 6: 1188–99.

Aristotle (2000) *Nicomachean Ethics*, Cambridge: Cambridge University Press.

Arkin, R. (2010) 'The Case for Ethical Autonomy in Unmanned Systems', *Journal of Military Ethics*, 9, 4: 332–41.

Banks, I.M. (1992) *The Use of Weapons*, London: Orbit.

Brandt, R.B. (1992) *Morality, Utilitarianism and Rights*, Cambridge: Cambridge University Press.

Cannon-Bowers, J.A. and Salas, E. (eds) (1998) *Making Decisions under Stress: Implications for Individual and Team Training*, Washington, D.C.: APA Books.

Cavanaugh, T.A. (2006) *Double Effect Reasoning: Doing Good and Avoiding Evil*, Oxford: Oxford University Press.

De Greef, T.E., Arciszewski, H.F.R., and Neerincx, M.A. (2010) 'Adaptive Automation Based on an Object-Oriented Task Model: Implementation and Evaluation in a Realistic C2 Environment,' *Journal of Cognitive Engineering and Decision Making*, 31: 152–82.

Dennett, D. (1997) 'When HAL Kills, Who's to Blame? Computer Ethics', in Stork, D.G. (ed.) *HAL's Legacy: 2001's Computer as Dream and Reality*, Cambridge, MA: MIT Press.

Finnis, J. and Boyle, J. Grisez (eds) (1988) *Nuclear Deterrence, Morality and Realism*, Oxford: Oxford University Press.

Hollnagel, E. and Woods, D.D. (2005) *Joint Cognitive Systems: Foundations of Cognitive Systems Engineering*, Boca Raton: Taylor and Francis.

Human Rights Watch (2012) *Losing Humanity: The Case against Killer Robots*, New York, www.hrw.org/reports/2012/11/19/losing-humanity (accessed 25 November 2013).

Kant, I. (1998) *Groundwork of the Metaphysics of Morals*, Cambridge: Cambridge University Press.

Mathias, A. (2004) 'The Responsibility Gap: Ascribing Responsibility for the Actions of Learning Automata,' *Ethics and Information Technology*, 6: 175–83.

McGinnis, J.O. (2010) 'Accelerating AI', *Northwestern University Law Review*, 104: 1253–70.

McMahan, J. (2009) *Killing in War*, Oxford: Oxford University Press.

Miller, C.A. and Parasuraman, R. (2007) 'Designing a Flexible Interaction between Humans and Automation: Delegation Interfaces for Supervisory Control', *Human Factors*, 49, 1: 57–75.

Rawls, J. (1999a) *A Theory of Justice*, Oxford: Oxford University Press.

Rawls, J. (1999b) *The Law of Peoples*, Cambridge, MA: Harvard University Press.

Sharkey, N. (2010) 'Saying "No" to Lethal Autonomous Targeting', *Journal of Military Ethics*, 9, 4: 369–83.

Sparrow, R. (2007) 'Killer Robots', *Journal of Applied Philosophy*, 24, 1: 62–77.

Strawser, B.J. (ed.) (2013) *Killing by Remote Control: The Ethics of an Unmanned Military*, Oxford: Oxford University Press.

United States Department of Defense (2011) *Unmanned Systems Integrated Roadmap FY 2011–2036*, Reference Number 11-S-3613, www.defenseinnovationmarketplace.mil/resources/UnmannedSystemsIntegratedRoadmapFY2011.pdf (accessed 25 November 2013).

Walzer, M. (2006) *Just and Unjust Wars: A Moral Argument with Historical Illustrations*, 4th edn, New York: Basic Books.

Conclusions

Precision strikes – the way forward

Wali Aslam and Regina Rauxloh

There can be little doubt that the use of precision strike capabilities will increase in the near future since they have already become a central element of modern warfare. In particular, the development of Unmanned Aerial Vehicles (UAVs) is a child of our time, reflecting not only rapid technological developments, especially in computer science and satellite technology, but also a change in attitude to large-scale military force. The current need to cut down on military spending during the global financial crisis, the (alleged) necessity to fight non-state actors abroad without invading the host country, the increasing reluctance to accept mass human rights atrocities abroad, but also to risk the lives of soldiers, put pressures on modern militaries to focus more on developing weapons systems to deal with these challenges. Indeed, the contributions by Shea and Franke elaborate on how a large-scale proliferation of the UAV capability has already begun, a development which is unlikely to be reversed.

This chapter will seek to draw conclusions from the variety of contributions to this book about how research on different aspects of the subject of precision strikes can be advanced. The contributions to this volume demonstrate that a significant amount of future research in this area, covering a broad spectrum, is necessary. The chapter will begin by looking at the specific areas which require further exploration. It will also make the case for a need to conduct cross-disciplinary research ranging from psychology to political science and from mechanical engineering to sociology if the transformative impact of the developments in these fields is to be fully understood. Finally, it will be argued that any meaningful research in this area will require a much stronger engagement between different stakeholders such as academics, policy makers, military decision makers, NGOs, and the media.

Further research

The acquisition by many states of precision strike capabilities has significantly changed the way wars are waged. As far as UAVs are concerned, according to Barrinha and da Vinha they are 'no longer mere complementary pieces of a larger military puzzle; they are its core: the starting point from which everything else is enabled'. Similarly, Franke highlights how this capability is the new military 'must have' and its symbolic value with regard to military modernisation must be understood if one is to fully comprehend why such a large number of states are attempting to acquire it. It is therefore imperative that we have a thorough knowledge and understanding of the implications of acquiring and deploying UAVs and other precision strike capabilities. While this book has dealt with a number of these implications, several key questions which require further research have been identified throughout the volume.

Anticipating and shaping future developments

Most of the debate around precision strike capabilities and the use of drones focuses on the practices of Israel and, above all, the United States. As Franke's chapter has shown, however, the use of UAVs is likely to spread to all major military powers. Thus, the debate about precision strike capabilities should not only discuss their current use but also their likely development in the future. The research by Dyson shows very clearly how a range of social, cultural, and political factors play a role in determining the differentiated acquisition and use of precision strike capabilities by the West European Great Powers. However, this research requires broadening to other military powers in the international system as the current studies predominantly focus on the US, Israel, and Europe (more broadly). For example, a study concerning Chinese, Indian, or Pakistani activities in this regard would be crucial in giving an indication of the changing political and security dynamics in South Asia. There is a major gap in the literature concerning a number of significant cases that could be studied from the conceptual perspectives addressed in this volume. Such a research would help develop a more fulsome understanding of the factors which facilitate successful reform in the area of precision strike capabilities.

Further research is also required to deepen our understanding of the relative importance of material and ideational variables in the development of precision strike capabilities. As Franke notes, strategic culture can provide a useful analytical framework for understanding the global diffusion of precision strike technology by helping to illuminate the impact of ideational factors on military emulation. On the other hand, as Dyson's contribution demonstrates, realist examinations provide a stronger analytical leverage in understanding the process (and outcome) of advancement in the field of precision strikes. Future theory testing in comparative case study research would deepen our understanding of the utility of these theoretical frameworks and should focus on the potential for dialogue between cultural and realist approaches.

Precision strikes and decision making models

One key lesson this volume has highlighted is that while advances in military technology decrease the role of human actors, humans will always be involved in the process to some extent. Therefore another key area that needs further exploration is the process of decision making concerning technologies where the decision making is shared between human beings and machines as is the case with UAVs. Banks and Dhami's chapter explains the various psychological factors affecting different actors at different levels of decision making. They highlight how studies concerning the use of UAVs focus on examining rational and analytic thinking. However, such a process of decision making is not always possible, especially when the operator is subject to information overload and may have to rely on heuristic mechanisms of decision making. Hence, in order to fully comprehend how decisions are made, it is imperative to understand the element of 'quasi-rationality' as highlighted by these authors. As such the question of how decision making can be shared with – or even replaced by – non-human actors is not just a question of engineering but rather a problem which belongs to the discipline of psychology.

International law

Remotely controlled weapon systems facilitate a very different use of armed force from traditional warfare, which might make it necessary to amend the current law of armed conflict. In the fight against international terrorism the traditional definitions of combatants and civilians do not any longer serve their purpose in identifying legitimate targets (Swiney, 2005). Much more work is therefore needed to establish whether the current distinction needs to be adjusted to include new categories of fighters and to decide what their respective rights are, for example as prisoners of war. In addition, the principle of proportionality requires that any collateral damage must be proportionate to the anticipated military advantage of the attack. While much literature discusses the exact number of casualties and the reasons why there is so much disagreement about this,[1] more research is needed to evaluate the military advantage of each drone strike. This includes two essential questions. First, how reliable is intelligence gathered at the moment in order to make a clear judgement as to the expected military advantage of each strike. Second, if the anticipated military gain is a reduction of the risk of a future terrorist attack, how can this vague advantage be compared with the concrete loss of human life and other damage (e.g. to the environment or cultural property)? International legal regimes will have to engage with these questions in order for there to be a meaningful examination of these advancements in technology.

Furthermore, as Ruettershoff's contribution makes clear, the possibility of launching attacks from (but not against) a foreign state raises the crucial question of how precision strikes will influence our current notion of territorial sovereignty. The question of the use of force against non-state actors was raised during the Cold War where the superpowers conducted proxy warfare to weaken their opponents, but was never satisfactorily clarified in international law.[2] Both the growing internationalisation of terrorism and the possibility of remote controlled military attacks make this question more important than ever. This leads to a number of questions. How can we protect smaller states from being forced to give consent to military action by another state and thus risk becoming an 'open battlefield'? Are we observing an erosion of the principles of non-intervention, equality of sovereign states, and the national control of territory? Would these developments be conducive to promoting order in the international community or would they be a recipe for chaos as discussed by Power, Alison, and Ralph? A more specific question which is not limited to, but is highlighted by the discussions in this volume is whether the international community should create a legal regime regulating forms of anti-terrorist actions through precision strikes, such as targeted killing. It has been suggested that the Missile Technology Control Regime can serve as a starting point for that purpose. The question of targeted killing, however, is not merely a legal one. For a meaningful discussion of this topic the input of people from many different disciplines, such as moral theorists, national security experts, and political scientists is necessary.

All of these questions should be addressed not in a piecemeal approach but in a holistic review of the scope and role of international law in relation to new forms of armed conflict. This reassessment must take place in the wider context of the role and the utility of the use of military force. The discussion needs to review theoretical underpinnings such as the concept of 'risk management warfare', as explored by Barrinha and da Vinha. The extent to which a redistribution of risk by more technologically advanced states is acceptable to the international community requires exploration. A more practical question which follows from this discussion is whether the development and proliferation of precision strike capabilities should face much tighter regulations despite the likely resistance to such a move.

Actors of warfare

A further question which requires much more thorough examination is the political accountability, legitimacy, and compliance with international law of the small-scale, clandestine use of force by secret services like the CIA. The dispute about the exact numbers of casualties caused by American UAVs (as discussed by Waddington and Woods) demonstrates the underlying political problems concerning a lack of transparency in such covert operations. The White House has recently begun to recognise that the UAV campaign can lose legitimacy (despite its successes) if it is shrouded in secrecy. The attempts by various Obama administration officials (such as Eric Holder and John Brennan), and by the President himself, to enhance the transparency of the campaign show a gradual move towards openness. It is likely that this development will continue in the future. The US is cognisant of the fact that the world is moving towards a multipolar system and, in the context of the shift of US strategic interests to Asia, the UK and France are – and must become – increasingly involved in projecting military power within Europe's geopolitical neighbourhood, particularly Africa. Furthermore, with the arrival of new great powers (such as China and India), it is likely that the US will find itself more constrained from acting in a unilateral and aggressive manner. This will necessitate the US's seeking greater legitimacy for its action with a view to increase increasing its influence. However, in order to gain legitimacy, the US will have to justify its actions in front of national and international audiences, making transparency an imperative.

Tactical advantages and disadvantages

The contribution by Power, Alison, and Ralph notes the current lack of understanding of short, medium, and long-term effects of precision strike technologies, such as UAVs. Indeed, although there is global interest in acquiring and using precision strike capabilities, there is very limited literature on the actual successes of these technologies. It is often claimed that such technologies are more able and likely to reduce civilian casualties, shorten (or even avoid) full-blown wars and, in general, make the use of force more humane (Strawser, 2010). But in order to test this claim we need to gain a much clearer picture as to how precise and capable precision strike capabilities actually are. Waddington's point that the proficiencies of these systems are often overestimated may be a key area for further exploration.

It has also been argued by Leveringhaus and de Greef that with a number of important caveats the use of automatic and autonomous weapons may be preferable to traditional modes of war because the machines are immune to psychological stress and fear and thus less prone to commit judgment errors. Some may argue that another important advantage is that they do not commit war crimes such as using sexual violence as a weapon of war or committing torture. This may be a valid argument but needs further scrutiny.

The debate on the use of UAVs is currently focussed on counterterrorism activities (Aslam, 2012) and operations of targeted killing (Braun and Brunstetter, 2013). However, the humanitarian potential of UAVs, where they can be employed by organisations such as the UN for humanitarian intervention and peacekeeping operations, needs further enquiry. While many academics are very critical (and sceptical) of the current use of UAVs (see, for example, Waddington's chapter), Wiesner suggests that this technology can play a very useful and important role in humanitarian interventions and the strengthening and improvement of the doctrine of the Responsibility to Protect. Wiesner suggests that UAVs can play a vital role in intelligence, surveillance, and reconnaissance (ISR) operations which can be crucial for peacekeeping and humanitarian operations. However, the exact roles UAVs can play in gathering information about a conflict, whether it is to help the UN to decide about necessary actions, monitor a ceasefire, or provide evidence for later international criminal trials, needs to be further researched and developed. The potential use of remote controlled armed weapon systems by the UN as part of peacekeeping operations and humanitarian interventions also needs to be discussed.

Strategic advantages and disadvantages

The utility of precision strike counter-terrorist action also requires analysis in relation to understanding its long-term impact on regions and states at the receiving end, as well as the impact on the national security of the aggressor state. We require a clearer picture not only about how individual victims are affected, but also about the overall impact on communities living under the threat of drone attacks. How does it impact local economic and political life given that there is likely to be an increase in the frequency of such strikes (through both manned and remotely piloted aircraft?) Equally important is the question of how these technologies might have longer-term negative consequences if they end up increasing recruitment to terrorist groups to avenge civilian casualties. What effects do precision technologies have on the modes and localities of terrorist operations? To what extent do military actions (relying on such technologies) strengthen the ideological claims of terrorist networks?

Another area that requires further research is the politicisation of the use of these technologies by different groups and political parties who are at the receiving end of this advancement in technology. There have been instances which demonstrate that this is happening and that counterterrorism and counterinsurgency operations are being politicised in targeted states by certain groups for their narrow agendas. The contribution by Aslam demonstrates a major gap in literature concerning research involving the 'other side' – those at the receiving end of these advancements in military technologies. One potential area of research is to examine how the use of UAVs can have political consequences on the ground where different actors may appropriate (or 'hijack') and politicise the drone casualties to accomplish their own agendas. As Aslam shows, the 'political appropriation' of the drone casualties is helping radicalise Pakistani youth. Such unforeseen consequences of the use of UAVs would have to be considered for a comprehensive evaluation of the use of such technologies.

These by-products of counter-terrorist strikes affect many different countries, not only the one which is conducting the precision strikes.

Barrinha and da Vinha point out that the solutions to security concerns are increasingly sought through advanced technology rather than in social, economic, political, or ideological dialogue. According to Barrinha and da Vinha, while new technology 'allows for radically shrinking the "find-fix-finish" loop, contributing to greater efficiency in eliminating targets', other (more effective as well as more humanitarian) solutions might be overlooked.

Over-reliance on technological advancement may lead one to think that such developments can serve as a replacement for the more complicated business of nation building. The United States increasingly started to focus on revolutionising its conventional military technology in the post-Vietnam era when it was reluctant to put boots-on-the-ground in protracted low-intensity conflicts. Those technologies were seen as the future of warfare during the 1980s and 1990s. However, the invasions of Afghanistan (2001) and Iraq (2003) made clear the limits of such technologies, and showed that while they may serve some initial purposes of dislodging an enemy regime from power, they cannot serve as an alternative for a more comprehensive policy of nation-building. Interestingly, the recent American involvement in Iraq crystallised the problems that may emerge when the intervening power is reluctant to dispatch soldiers to the theatre of war.

The recent post-Iraq focus on precision strikes (and especially the use of drones) looks like a return to the post-Vietnam era with its concern for investing in transformative technologies. However, history instructs that future mistakes can be avoided by remembering the limits of such technologies in the past. At times, the intervening states may have to return to the deployment of traditional land-force that is necessary to stabilise the region after the achievement of the initial goals, such as regime change. Furthermore, such actors will also have to be ready to deal with the longer-term social, economic, and political problems in order to deal with the root-causes of the conflicts. Hence, a key lesson of the Western involvement in Iraq and Afghanistan is that technology cannot solve all the problems that emerge after an intervention and a misjudged belief in the power of technology to be the magical solution is most likely to lead to failure.

Cross-disciplinary research

The discussion concerning the advancement in precision strike capabilities covers many different areas, including engineering, international law, international relations, political science, psychology, and sociology. There are questions which must not be conflated, such as the question of choice of weapon and the choice of using force in the first place, but at the same time all of these questions are closely interrelated. No one discipline can claim to be able to address this field comprehensively and indeed the discussions surrounding precision strikes highlight how the boundaries between different disciplines are increasingly being eroded. Regrettably, however, as Wiesner points out '[w]hat is interesting, though, is to see that within the different areas each discourse is rather homogenous whereas between those fields only little exchange of arguments seems to exist'.

Although contributors to this collection deepen the insight by shining light on the topic from so many different angles, the individual chapters are still written from the viewpoint of their respective discipline. Yet, this book has shown that each of these areas is dependent on the other. For example, the discussion of collateral damage (Woods) heavily relies on media reports (Waddington). On the other hand, the classification of combatants (and victims) should be based on scientific definition (Ruettershoff), rather than the need for political interests to create and maintain legitimacy (Aslam).

This book demonstrates how drawing together different research areas is not only fruitful but necessary to develop a clear understanding of all the facets of this timely topic. But above and beyond the need for cross-disciplinary research the current chapter also makes the case for a dialogue between scholars, technical experts (such as scientists and engineers), policymakers, activists, and the media. The advancement in scientific technologies cannot be left in the hands of scientists only but requires a dialogue among all stakeholders. The academic scholars working in this field have a duty to take into account the interest of all affected parties whether they are politicians, victims, or security-related actors, such as the personnel of private military companies. This will stop the discussion from being limited to the proverbial ivory tower of academia and will enable wider society to use the research for practical purposes. A good example for the need for co-operation of different disciplines and stakeholders is the proposal put forward by Leveringhaus and de Greef: the development of an e-Partnership between human operators and artificial agents will only be possible through close cross-disciplinary research and cooperation between scientists, human science scholars, military specialists, and engineers.

The workshop organised by cii – The Centre for International Intervention at the University of Surrey – in the summer of 2012 and the RUSI report[3] which grew out of the findings of these sessions was an important step in this direction. The cross disciplinary research of this book is another step in encouraging cross fertilisation not only between different scholarly disciplines but also between academics, journalists, scientists, practitioners, policy makers, and the media.

It is hoped that this book will initiate a dialogue between practitioners and academics, which until now unfortunately remains quite limited. There are various political and practical issues that create obstacles in the path of such a dialogue in this field. Some impediments relate to academics' concerns that most practitioners are mostly concerned with superficial cost-benefit analyses without much interest in the wider, longer-term perspective. This book has, however, endeavoured to ensure that this platform is not misused for pushing forward a biased stance either way and showed that unbiased analyses of precision strike technologies are possible. Practitioners, on the other hand, consider academic writings to be too concerned with abstract theoretical discussions with limited advice to offer on day-to-day issues. However, the contributions in this book demonstrate that this does not have to be the case. Academics can offer many valuable insights that offer guidance on day-to-day issues concerning precision strikes, such as the mental health of drone or aircraft pilots as well as the on-the-ground consequences of their use.

The initiation of dialogue between practitioners and academics is vital if advances in technology are to be used for the wider benefit of humanity. This book takes a small step toward this goal. The various contributions have illustrated not only how many areas are affected by advancements in precision strike technologies but also how under-researched the various impacts are. We conclude that the decisions of policy makers and military personnel are far too important to be made without the direct input of a wide array of disciplines. The obstacles to dialogue between academia and policy makers may be significant, but it is essential that these disciplines make their voices heard.

Notes

1 For example, see the work of the Bureau of Investigative Journalism in that regard at their website: www.thebureauinvestigates.com.

2 Case Concerning the Military and Paramilitary Activities in and against Nicaragua (*Nicaragua v. United States of America*), ICJ Reports 1986; Armed Activities on the Territory of the Congo (*Democratic Republic of the Congo v. Uganda*), ICJ Reports 2005, especially separate opinions of Judge Kooijmans and Judge Simma.

3 www.rusi.org/publications/whitehallreports/ref:O51509D843E399.

References

Aslam, W. (2012) 'Understanding the "Pak" in "AfPak": The Obama Administration's Security Policy towards Pakistan at the Mid-term', *Journal of Policing, Intelligence and Counter Terrorism*, 7(1): 2–21.

Braun, M. and Brunstetter, D.R. (2013) 'Rethinking the Criterion for Assessing CIA-Targeted Killings: Drones, Proportionality and Jus ad vim', *Journal of Military Ethics*, 12(4): 304–24.

Strawser, J. (2010) 'Moral Predators: The Duty to Employ Uninhabited Aerial Vehicles', *Journal of Military Ethics*, 9(4): 342–68.

Swiney, G. (2005) 'Saving Lives: The Principle of Distinction and the Realities of Modern War', *International Lawyer*, 39: 733–58.

Index